Java 大数据分析(影印版)
Big Data Analytics with Java

Rajat Methta 著

南京　东南大学出版社

图书在版编目(CIP)数据

Java 大数据分析：英文/(美)拉贾特·梅塔(Rajat Mehta)著. —影印本. —南京：东南大学出版社，2019.3

书名原文：Big Data Analytics with Java

ISBN 978-7-5641-8287-8

Ⅰ.①J… Ⅱ.①拉… Ⅲ.①JAVA 语言-程序设计-英文 Ⅳ.①TP312.8

中国版本图书馆 CIP 数据核字(2019)第 025356 号

图字：10-2018-492 号

© 2017 by PACKT Publishing Ltd.

Reprint of the English Edition, jointly published by PACKT Publishing Ltd and Southeast University Press, 2019. Authorized reprint of the original English edition, 2017 PACKT Publishing Ltd, the owner of all rights to publish and sell the same.

All rights reserved including the rights of reproduction in whole or in part in any form.

英文原版由 PACKT Publishing Ltd 出版 2017。

英文影印版由东南大学出版社出版 2019。此影印版的出版和销售得到出版权和销售权的所有者 —— PACKT Publishing Ltd 的许可。

版权所有，未得书面许可，本书的任何部分和全部不得以任何形式重制。

Java 大数据分析（影印版）

出版发行：	东南大学出版社
地　　址：	南京四牌楼 2 号　邮编：210096
出 版 人：	江建中
网　　址：	http://www.seupress.com
电子邮件：	press@seupress.com
印　　刷：	常州市武进第三印刷有限公司
开　　本：	787 毫米×980 毫米　16 开本
印　　张：	26
字　　数：	509 千字
版　　次：	2019 年 3 月第 1 版
印　　次：	2019 年 3 月第 1 次印刷
书　　号：	ISBN 978-7-5641-8287-8
定　　价：	98.00 元

本社图书若有印装质量问题，请直接与营销部联系。电话(传真)：025-83791830

Credits

Author
Rajat Mehta

Reviewers
Dave Wentzel
Roberto Casati

Commissioning Editor
Veena Pagare

Acquisition Editor
Chandan Kumar

Content Development Editor
Deepti Thore

Technical Editors
Jovita Alva
Sneha Hanchate

Copy Editors
Safis Editing
Laxmi Subramanian

Project Coordinator
Shweta H Birwatkar

Proofreader
Safis Editing

Indexer
Pratik Shirodkar

Graphics
Tania Dutta

Production Coordinator
Shantanu N. Zagade

Cover Work
Shantanu N. Zagade

About the Author

Rajat Mehta is a VP (technical architect) in technology at JP Morgan Chase in New York. He is a Sun certified Java developer and has worked on Java-related technologies for more than 16 years. His current role for the past few years heavily involves the use of a big data stack and running analytics on it. He is also a contributor to various open source projects that are available on his GitHub repository, and is also a frequent writer for dev magazines.

About the Reviewers

Dave Wentzel is the CTO of Capax Global, a data consultancy specializing in SQL Server, cloud, IoT, data science, and Hadoop technologies. Dave helps customers with data modernization projects. For years, Dave worked at big independent software vendors, dealing with the scalability limitations of traditional relational databases. With the advent of Hadoop and big data technologies everything changed. Things that were impossible to do with data were suddenly within reach.

Before joining Capax, Dave worked at Microsoft, assisting customers with big data solutions on Azure. Success for Dave is solving challenging problems at companies he respects, with talented people who he admires.

Roberto Casati is a certified enterprise architect working in the financial services market. Roberto lives in Milan, Italy, with his wife, their daughter, and a dog.

In a former life, after graduating in engineering, he worked as a Java developer, Java architect, and presales architect for the most important telecommunications, travel, and financial services companies.

His interests and passions include data science, artificial intelligence, technology, and food.

www.PacktPub.com

eBooks, discount offers, and more

Did you know that Packt offers eBook versions of every book published, with PDF and ePub files available? You can upgrade to the eBook version at `www.PacktPub.com` and as a print book customer, you are entitled to a discount on the eBook copy. Get in touch with us at `customercare@packtpub.com` for more details.

At `www.PacktPub.com`, you can also read a collection of free technical articles, sign up for a range of free newsletters and receive exclusive discounts and offers on Packt books and eBooks.

`https://www.packtpub.com/mapt`

Get the most in-demand software skills with Mapt. Mapt gives you full access to all Packt books and video courses, as well as industry-leading tools to help you plan your personal development and advance your career.

Why subscribe?

- Fully searchable across every book published by Packt
- Copy and paste, print, and bookmark content
- On demand and accessible via a web browser

Customer Feedback

Thanks for purchasing this Packt book. At Packt, quality is at the heart of our editorial process. To help us improve, please leave us an honest review on this book's Amazon page at https://www.amazon.com/dp/1787288986.

If you'd like to join our team of regular reviewers, you can e-mail us at customerreviews@packtpub.com. We award our regular reviewers with free eBooks and videos in exchange for their valuable feedback. Help us be relentless in improving our products!

This book is dedicated to my mother Kanchan, my wife Harpreet, my daughter Meher, my father Ashwini and my son Vivaan.

Table of Contents

Preface	**vii**
Chapter 1: Big Data Analytics with Java	**1**
Why data analytics on big data?	3
Big data for analytics	3
Big data – a bigger pay package for Java developers	4
Basics of Hadoop – a Java sub-project	4
Distributed computing on Hadoop	8
HDFS concepts	8
Design and architecture of HDFS	9
Main components of HDFS	10
HDFS simple commands	11
Apache Spark	12
Concepts	12
Transformations	13
Actions	14
Spark Java API	15
Spark samples using Java 8	16
Loading data	16
Data operations – cleansing and munging	17
Analyzing data – count, projection, grouping, aggregation, and max/min	17
Actions on RDDs	19
Paired RDDs	20
Saving data	23
Collecting and printing results	23
Executing Spark programs on Hadoop	23
Apache Spark sub-projects	24
Spark machine learning modules	25
Mahout – a popular Java ML library	26
Deeplearning4j – a deep learning library	26
Summary	**27**

Table of Contents

Chapter 2: First Steps in Data Analysis — 29
Datasets — 29
Data cleaning and munging — 31
Basic analysis of data with Spark SQL — 33
Building SparkConf and context — 34
Dataframe and datasets — 34
Load and parse data — 35
Analyzing data – the Spark-SQL way — 35
Spark SQL for data exploration and analytics — 43
Market basket analysis – Apriori algorithm — 43
Implementation of the Apriori algorithm in Apache Spark — 51
Efficient market basket analysis using FP-Growth algorithm — 54
Running FP-Growth on Apache Spark — 66
Summary — 68

Chapter 3: Data Visualization — 69
Data visualization with Java JFreeChart — 69
Using charts in big data analytics — 70
Time Series chart — 71
All India seasonal and annual average temperature series dataset — 71
Simple single Time Series chart — 72
Multiple Time Series on a single chart window — 75
Bar charts — 77
Histograms — 80
When would you use a histogram? — 81
How to make histograms using JFreeChart? — 81
Line charts — 82
Scatter plots — 84
Box plots — 88
Advanced visualization technique — 95
Prefuse — 95
IVTK Graph toolkit — 96
Other libraries — 96
Summary — 96

Chapter 4: Basics of Machine Learning — 99
What is machine learning? — 100
Real-life examples of machine learning — 100
Type of machine learning — 102
A small sample case study of supervised and unsupervised learning — 106
Steps for machine learning problems — 107
Choosing the machine learning model — 110
What are the feature types that can be extracted from the datasets? — 111
How do you select the best features to train your models? — 114

How do you run machine learning analytics on big data?	119
Getting and preparing data in Hadoop	120
Training and storing models on big data	123
Apache Spark machine learning API	125
Summary	**127**

Chapter 5: Regression on Big Data — 129

Linear regression — 130
What is simple linear regression? — 132
Where is linear regression used? — 135
Logistic regression — 143
Which mathematical functions does logistic regression use? — 144
Where is logistic regression used? — 146
Predicting heart disease using logistic regression — 147
Summary — 153

Chapter 6: Naive Bayes and Sentiment Analysis — 155

Conditional probability — 156
Bayes theorem — 157
Naive Bayes algorithm — 159
Advantages of Naive Bayes — 160
Disadvantages of Naive Bayes — 161
Sentimental analysis — 162
Concepts for sentimental analysis — 162
Tokenization — 163
Stop words removal — 163
Stemming — 164
N-grams — 165
Term presence and Term Frequency — 165
TF-IDF — 166
Bag of words — 168
Dataset — 168
Data exploration of text data — 169
Sentimental analysis on this dataset — 174
SVM or Support Vector Machine — 181
Summary — 183

Chapter 7: Decision Trees — 185

What is a decision tree? — 185
Building a decision tree — 188
Choosing the best features for splitting the datasets — 191
Dataset — 196
Data exploration — 197
Cleaning and munging the data — 201
Training and testing the model — 202
Summary — 209

Chapter 8: Ensembling on Big Data — 211
Ensembling — 212
Types of ensembling — 213
- Bagging — 213
- Boosting — 215
- Advantages and disadvantages of ensembling — 216
Random forests — 218
Gradient boosted trees (GBTs) — 219
- Classification problem and dataset used — 221
- Data exploration — 222
- Training and testing our random forest model — 230
- Training and testing our gradient boosted tree model — 236
Summary — 237

Chapter 9: Recommendation Systems — 239
Recommendation systems and their types — 240
Content-based recommendation systems — 242
Dataset — 248
Content-based recommender on MovieLens dataset — 249
Collaborative recommendation systems — 256
- Advantages — 257
- Disadvantages — 258
- Alternating least square – collaborative filtering — 258
Summary — 266

Chapter 10: Clustering and Customer Segmentation on Big Data — 267
Clustering — 268
Types of clustering — 270
- Hierarchical clustering — 270
- K-means clustering — 272
- Bisecting k-means clustering — 273
Customer segmentation — 275
Dataset — 276
Data exploration — 276
Clustering for customer segmentation — 280
- Changing the clustering algorithm — 287
- Summary — 288

Chapter 11: Massive Graphs on Big Data — 289
Refresher on graphs — 290
Representing graphs — 292
- Common terminology on graphs — 293
- Common algorithms on graphs — 294
- Plotting graphs — 295

Massive graphs on big data	**297**
Graph analytics	298
GraphFrames	300
Building a graph using GraphFrames	300
Graph analytics on airports and their flights	304
Datasets	305
Graph analytics on flights data	306
Summary	**319**
Chapter 12: Real-Time Analytics on Big Data	**321**
Real-time analytics	**322**
Big data stack for real-time analytics	324
Real-time SQL queries on big data	324
Real-time data ingestion and storage	325
Real-time data processing	325
Real-time SQL queries using Impala	326
Flight delay analysis using Impala	327
Apache Kafka	331
Spark Streaming	333
Trending videos	341
Summary	**352**
Chapter 13: Deep Learning Using Big Data	**353**
Introduction to neural networks	**354**
Perceptron	**356**
Problems with perceptrons	359
Sigmoid neuron	361
Multi-layer perceptrons	362
Accuracy of multi-layer perceptrons	364
Deep learning	**366**
Advantages and use cases of deep learning	366
Flower species classification using multi-Layer perceptrons	**367**
Deeplearning4j	**373**
Hand written digit recognizition using CNN	**374**
Diving into the code:	374
Summary	**383**
Index	**385**

Preface

Even as you read this content, there is a revolution happening behind the scenes in the field of big data. From every coffee that you pick up from a coffee store to everything you click or purchase online, almost every transaction, click, or choice of yours is getting analyzed. From this analysis, a lot of deductions are now being made to offer you new stuff and better choices according to your likes. These techniques and associated technologies are picking up so fast that as developers we all should be a part of this new wave in the field of software. This would allow us better prospects in our careers, as well as enhance our skill set to directly impact the business we work for.

Earlier technologies such as machine learning and artificial intelligence used to sit in the labs of many PhD students. But with the rise of big data, these technologies have gone mainstream now. So, using these technologies, you can now predict which advertisement the user is going to click on next, or which product they would like to buy, or it can also show whether the image of a tumor is cancerous or not. The opportunities here are vast. Big data in itself consists of a whole lot of technologies whether cluster computing frameworks such as Apache Spark or Tez or distributed filesystems such as HDFS and Amazon S3 or real-time SQL on underlying data using Impala or Spark SQL.

This book provides a lot of information on big data technologies, including machine learning, graph analytics, real-time analytics and an introductory chapter on deep learning as well. I have tried to cover both technical and conceptual aspects of these technologies. In doing so, I have used many real-world case studies to depict how these technologies can be used in real life. So this book will teach you how to run a fast algorithm on the transactional data available on an e-commerce site to figure out which items sell together, or how to run a page rank algorithm on a flight dataset to figure out the most important airports in a country based on air traffic. There are many content gems like these in the book for readers.

What this book covers

Chapter 1, *Big Data Analytics with Java,* starts with providing an introduction to the core concepts of Hadoop and provides information on its key components. In easy-to-understand explanations, it shows how the components fit together and gives simple examples on the usage of the core components HDFS and Apache Spark. This chapter also talks about the different sources of data that can put their data inside Hadoop, their compression formats, and the systems that are used to analyze that data.

Chapter 2, *First Steps in Data Analysis,* takes the first steps towards the field of analytics on big data. We start with a simple example covering basic statistical analytic steps, followed by two popular algorithms for building association rules using the Apriori Algorithm and the FP-Growth Algorithm. For all case studies, we have used realistic examples of an online e-commerce store to give insights to users as to how these algorithms can be used in the real world.

Chapter 3, *Data Visualization,* helps you to understand what different types of charts there are for data analysis, how to use them, and why. With this understanding, we can make better decisions when exploring our data. This chapter also contains lots of code samples to show the different types of charts built using Apache Spark and the JFreeChart library.

Chapter 4, *Basics of Machine Learning,* helps you to understand the basic theoretical concepts behind machine learning, such as what exactly is machine learning, how it is used, examples of its use in real life, and the different forms of machine learning. If you are new to the field of machine learning, or want to brush up your existing knowledge on it, this chapter is for you. Here I will also show how, as a developer, you should approach a machine learning problem, including topics on feature extraction, feature selection, model testing, model selection, and more.

Chapter 5, *Regression on Big Data,* explains how you can use linear regression to predict continuous values and how you can do binary classification using logistic regression. A real-world case study of house price evaluation based on the different features of the house is used to explain the concepts of linear regression. To explain the key concepts of logistic regression, a real-life case study of detecting heart disease in a patient based on different features is used.

Chapter 6, *Naive Bayes and Sentimental Analysis,* explains a probabilistic machine learning model called Naive Bayes and also briefly explains another popular model called the support vector machine. The chapter starts with basic concepts such as Bayes Theorem and then explains how these concepts are used in Naive Bayes. I then use the model to predict the sentiment whether positive or negative in a set of tweets from Twitter. The same case study is then re-run using the support vector machine model.

Chapter 7, *Decision Trees,* explains that decision trees are like flowcharts and can be programmatically built using concepts such as Entropy or Gini Impurity. The golden egg in this chapter is a case study that shows how we can predict whether a person's loan application will be approved or not using decision trees.

Chapter 8, *Ensembling on Big Data,* explains how ensembling plays a major role in improving the performance of the predictive results. I cover different concepts related to ensembling in this chapter, including techniques such as how multiple models can be joined together using bagging or boosting thereby enhancing the predictive outputs. We also cover the highly popular and accurate ensemble of models, random forests and gradient-boosted trees. Finally, we predict loan default by users in a dataset of a real-world Lending Club (a real online lending company) using these models.

Chapter 9, *Recommendation Systems,* covers the particular concept that has made machine learning so popular and it directly impacts business as well. In this chapter, we show what recommendation systems are, what they can do, and how they are built using machine learning. We cover both types of recommendation systems: content-based and collaborative, and also cover their good and bad points. Finally, we cover two case studies using the MovieLens dataset to show recommendations to users for movies that they might like to see.

Chapter 10, *Clustering and Customer Segmentation on Big Data,* speaks about clustering and how it can be used by a real-world e-commerce store to segment their customers based on how valuable they are. I have covered both k-Means clustering and bisecting k-Means clustering, and used both of them in the corresponding case study on customer segmentation.

Chapter 11, *Massive Graphs on Big Data,* covers an interesting topic, graph analytics. We start with a refresher on graphs, with basic concepts, and later go on to explore the different forms of analytics that can be run on the graphs, whether path-based analytics involving algorithms such as breadth-first search, or connectivity analytics involving degrees of connection. A real-world flight dataset is then used to explore the different forms of graph analytics, showing analytical concepts such as finding top airports using the page rank algorithm.

Chapter 12, Real-Time Analytics on Big Data, speaks about real-time analytics by first seeing a few examples of real-time analytics in the real world. We also learn about the products that are used to build real-time analytics system on top of big data. We particularly cover the concepts of Impala, Spark Streaming, and Apache Kafka. Finally, we cover two real-life case studies on how we can build trending videos from data that is generated in real-time, and also do sentiment analysis on tweets by depicting a Twitter-like scenario using Apache Kafka and Spark Streaming.

Chapter 13, Deep Learning Using Big Data, speaks about the wide range of applications that deep learning has in real life whether it's self-driving cars, disease detection, or speech recognition software. We start with the very basics of what a biological neural network is and how it is mimicked in an artificial neural network. We also cover a lot of the theory behind artificial neurons and finally cover a simple case study of flower species detection using a multi-layer perceptron. We conclude the chapter with a brief introduction to the Deeplearning4j library and also cover a case study on handwritten digit classification using convolution neural networks.

What you need for this book

There are a few things you will require to follow the examples in this book: a text editor (I use Sublime Text), internet access, admin rights to your machine to install applications and download sample code, and an IDE (I use Eclipse and IntelliJ).

You will also need other software such as Java, Maven, Apache Spark, Spark modules, the GraphFrames library, and the JFreeChart library. We mention the required software in the respective chapters.

You also need a good computer with a good RAM size, or you can also run the samples on Amazon AWS.

Who this book is for

If you already know some Java and understand the principles of big data, this book is for you. This book can be used by a developer who has mostly worked on web programming or any other field to switch into the world of analytics using machine learning on big data.

A good understanding of Java and SQL is required. Some understanding of technologies such as Apache Spark, basic graphs, and messaging will also be beneficial.

Conventions

In this book, you will find a number of styles of text that distinguish between different kinds of information. Here are some examples of these styles, and an explanation of their meaning.

A block of code is set as follows:

```
Dataset<Row> rowDS = spark.read().csv("data/loan_train.csv");
rowDS.createOrReplaceTempView("loans");
Dataset<Row> loanAmtDS = spark.sql("select _c6 from loans");
```

When we wish to draw your attention to a particular part of a code block, the relevant lines or items are set in bold:

```
Dataset<Row>data = spark.read().csv("data/heart_disease_data.csv");
    System.out.println("Number of Rows -->" + data.count());
```

> Warnings or important notes appear in a box like this.

> Tips and tricks appear like this.

Reader feedback

Feedback from our readers is always welcome. Let us know what you think about this book—what you liked or may have disliked. Reader feedback is important for us to develop titles that you really get the most out of.

To send us general feedback, simply send an e-mail to feedback@packtpub.com, and mention the book title via the subject of your message.

If there is a topic that you have expertise in and you are interested in either writing or contributing to a book, see our author guide on www.packtpub.com/authors.

If you have any questions, don't hesitate to look me up on LinkedIn via my profile https://www.linkedin.com/in/rajatm/, I will be more than glad to help a fellow software professional.

Customer support

Now that you are the proud owner of a Packt book, we have a number of things to help you to get the most from your purchase.

Downloading the example code

You can download the example code files for all Packt books you have purchased from your account at http://www.packtpub.com. If you purchased this book elsewhere, you can visit http://www.packtpub.com/support and register to have the files e-mailed directly to you.

You can download the code files by following these steps:

1. Log in or register to our website using your e-mail address and password.
2. Hover the mouse pointer on the **SUPPORT** tab at the top.
3. Click on **Code Downloads & Errata**.
4. Enter the name of the book in the **Search** box.
5. Select the book for which you're looking to download the code files.
6. Choose from the drop-down menu where you purchased this book from.
7. Click on **Code Download**.

You can also download the code files by clicking on the **Code Files** button on the book's webpage at the Packt Publishing website. This page can be accessed by entering the book's name in the **Search** box. Please note that you need to be logged in to your Packt account.

Once the file is downloaded, please make sure that you unzip or extract the folder using the latest version of:

- WinRAR / 7-Zip for Windows
- Zipeg / iZip / UnRarX for Mac
- 7-Zip / PeaZip for Linux

The code bundle for the book is also hosted on GitHub at https://github.com/PacktPublishing/Big-Data-Analytics-with-Java. We also have other code bundles from our rich catalog of books and videos available at https://github.com/PacktPublishing/. Check them out!

Downloading the color images of this book

We also provide you with a PDF file that has color images of the screenshots/diagrams used in this book. The color images will help you better understand the changes in the output. You can download this file from https://www.packtpub.com/sites/default/files/downloads/BigDataAnalyticswithJava_ColorImages.pdf.

Errata

Although we have taken every care to ensure the accuracy of our content, mistakes do happen. If you find a mistake in one of our books—maybe a mistake in the text or the code—we would be grateful if you would report this to us. By doing so, you can save other readers from frustration and help us improve subsequent versions of this book. If you find any errata, please report them by visiting http://www.packtpub.com/submit-errata, selecting your book, clicking on the **errata submission form** link, and entering the details of your errata. Once your errata are verified, your submission will be accepted and the errata will be uploaded on our website, or added to any list of existing errata, under the Errata section of that title. Any existing errata can be viewed by selecting your title from http://www.packtpub.com/support.

Piracy

Piracy of copyright material on the Internet is an ongoing problem across all media. At Packt, we take the protection of our copyright and licenses very seriously. If you come across any illegal copies of our works, in any form, on the Internet, please provide us with the location address or website name immediately so that we can pursue a remedy.

Please contact us at copyright@packtpub.com with a link to the suspected pirated material.

We appreciate your help in protecting our authors, and our ability to bring you valuable content.

Questions

You can contact us at questions@packtpub.com if you are having a problem with any aspect of the book, and we will do our best to address it.

1
Big Data Analytics with Java

Big data is no more just a buzz word. In almost all the industries, whether it is healthcare, finance, insurance, and so on, it is heavily used these days. There was a time when all the data that was used in an organization was what was present in their relational databases. All the other kinds of data, for example, data present in the `log` files were all usually discarded. This discarded data could be extremely useful though, as it can contain information that can help to do different forms of analysis, for example, `log` files data can tell about patterns of user interaction with a particular website. Big data helps store all these kinds of data, whether structured or unstructured. Thus, all the `log` files, videos, and so on can be stored in big data storage. Since almost everything can be dumped into big data whether they are `log` files or data collected via sensors or mobile phones, the amount of data usage has exploded within the last few years.

Three Vs define big data and they are volume, variety and velocity. As the name suggests, big data is a huge amount of data that can run into terabytes if not peta bytes of volume of storage. In fact, the size is so humongous that ordinary relational databases are not capable of handling such large volumes of data. Apart from data size, big data can be of any type of data be it the pictures that you took in the 20 years or the spatial data that a satellite sends, which can be of any type, be it text or in the form of images. Any type of data can be dumped into the big data storage and analyzed. Since the data is so huge it cannot fit on a single machine and hence it is stored on a group of machines. Many programs can be run in parallel on these machines and hence the speed or velocity of computation on big data. As the quantity of this data is very high, very insightful deductions can now be made from the data. Some of the use cases where big data is used are:

- In the case of an e-commerce store, based on a user's purchase history and likes, new set of products can be recommended to the users, thereby increasing the sales of the site

- Customers can be segmented into different groups for an e-commerce site and can then be presented with different marketing strategies
- On any site, customers can be presented with ads they might be most likely to click on
- Any regular ETL-like work (for example, as in finance or healthcare, and so on.) can be easily loaded into the big data stack and computed in parallel on several machines
- Trending videos, products, music, and so on that you see on various sites are all built using analytics on big data

Up until few years back, big data was mostly batch. Therefore, any analytics job that was run on big data was run in a batch mode usually using MapReduce programs, and the job would run for hours if not for days and would then compute the output. With the creation of the cluster computing framework, Apache Spark, a lot of these batch computations that took lot of time earlier have tremendously improved now.

Big data is not just Apache Spark. It is an ecosystem of various products such as Hive, Apache Spark, HDFS, and so on. We will cover these in the upcoming sections.

This book is dedicated to analytics on big data using Java. In this book, we will be covering various techniques and algorithms that can be used to analyze our big data.

In this chapter, we will cover:

- General details about what big data is all about
- An overview of the big data stack—Hadoop, HDFS, Apache Spark
- We will cover some simple HDFS commands and their usage
- We will provide an introduction to the core Spark API of RDDs using a few examples of its actions and transformations using Java
- We will also cover a general introduction on Spark packages such as MLlib, and compare them with other libraries such as Apache Mahout
- Finally, we will give a general description of data compression formats such as Avro and Parquet that are used in the big data world

Why data analytics on big data?

Relational databases are suitable for real-time CRUD operations such as order capture in e-commerce stores but they are not suitable for certain use cases for which big data is used. The data that is stored in relational databases is structured only but in big data stack (read Hadoop) both structured and unstructured data can be stored. Apart from this, the quantity of data that can be stored and parallelly processed in big data is massive. Facebook stores close to a tera byte of data in its big data stack on a daily basis. Thus, mostly in places where we need real-time CRUD operations on data, we can still continue to use relational databases, but in other places where we need to store and analyze almost any kind of data (whether `log` files, video files, web access logs, images, and so on.), we should use Hadoop (that is, big data).

Since analytics run on Hadoop, it runs on top of massive amounts of data; it is thereby a no brainer that deductions made from this are way more different than can be made from small amounts of data. As we all know, analytic results from large data amounts beat any fancy algorithm results. Also you can run all kinds of analytics on this data whether it be stream processing, predictive analytics, or real-time analytics.

The data on top of Hadoop is parallelly processed on multiple nodes. Hence the processing is very fast and the results are parallelly computed and combined.

Big data for analytics

Let's take a look at the following diagram to see what kinds of data can be stored in big data:

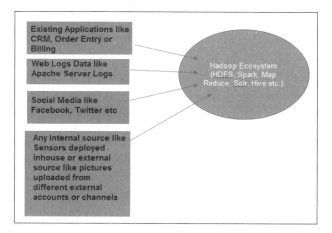

As you can see, the data from varied sources and of varied kinds can be dumped into Hadoop and later analyzed. As seen in the preceding image there could be many existing applications that could serve as sources of data whether providing CRM data, log data, or any other kind of data (for example, orders generated online or audit history of purchase orders from existing web order entry applications). Also as seen in the image, data can also be collected from social media or web logs of HTTP servers like Apache or any internal source like sensors deployed in a house or in the office, or external source like customers' mobile devices, messaging applications such as messengers and so on.

Big data – a bigger pay package for Java developers

Java is a natural fit for big data. All the big data tools support Java. In fact, some of the core modules are written in Java only, for example, Hadoop is written in Java. Learning some of the big data tools is no different than learning a new API for Java developers. So, putting big data skills in their skillset is a healthy addition for all the Java developers.

Mostly, Python and R language are hot in the field of data science mainly because of the ease of use and the availability of great libraries such as `scikit-learn`. But, Java, on the other hand has picked up greatly due to big data. On the big data side, there is availability of good software on the Java stack that can be readily used for applying regular analytics or predictive analytics using machine learning libraries.

Learning a combination of big data and analytics on big data would get you closer to apps that make a real impact on business and hence they command a good pay too.

Basics of Hadoop – a Java sub-project

Hadoop is a free, Java-based programming framework that supports the processing of these large datasets in a distributed computing environment. It is part of the Apache Software Foundation and was donated by Yahoo! It can be easily installed on a cluster of standard machines. Different computing jobs can then be parallelly run on these machines for faster performance. Hadoop has become very successful in companies to store all of their massive data in one system and perform analysis on this data. Hadoop runs in a master/slave architecture. The master controls the running of the entire distributed computing stack.

Some of the main features of Hadoop are:

Feature name	Feature description
Failover support	If one or more slave machines go down, the task is transferred to another workable machine by the master
Horizontal scalability	Just by adding a new machine, it comes within the network of the Hadoop framework and becomes part of the Hadoop ecosystem
Lower cost	Hadoop runs on cheap commodity hardware and is much cheaper than the costly large data solutions of other companies. For example some bigger firms have large data warehouse implementations such as Oracle Exadata or Teradata. These also let you store and analyze huge amounts of data but their hardware and software both are expensive and require more maintenance. Hadoop on the other hand installs on commodity hardware and its software is open sourced.
Data locality	This is one of the most important features of Hadoop and is the reason why Hadoop is so fast. Any processing of large data is done on the same machine on which the data resides. This way, there is no time and bandwidth lost in the transferring of data.

There is an entire ecosystem of software that is built around Hadoop. Take a look at the following diagram to visualize the Hadoop ecosystem:

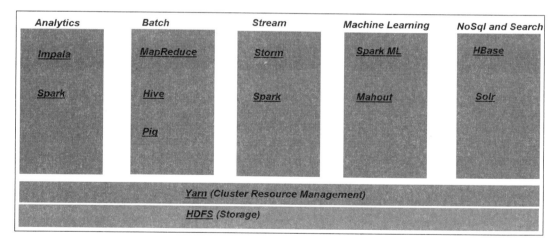

As you can see in the preceding diagram, for different criteria we have a different set of products. The main categories of the products that big data has are shown as follows:

- **Analytical products**: The whole purpose of this big data usage is an ability to analyze and make use of this extensive data. For example, if you have click stream data lying in the HDFS storage of big data and you want to find out the users with maximum hits or users who made the most number of purchases, or based on the transaction history of users you want to figure out the best recommendations for your users, there are some popular products that help us to analyze this data to figure out these details. Some of these popular products are Apache Spark and Impala. These products are sophisticated enough to extract data from the distributed machines of big data storage and to transform and manipulate it to make it useful.

- **Batch products**: in the initial stages when it came into picture, the word "big data" was synonymous with batch processing. So you had jobs that ran on this massive data for hours and hours cleaning and extracting the data to probably build useful reports for the users. As such, the initial set of products that shipped with Hadoop itself included "MapReduce", which is a parallel computing batch framework. Over time, more sophisticated products appeared such as Apache Spark, which also a cluster computing framework but is comparatively faster than MapReduce, but still in actuality they are batch only.

- **Streamlining**: This category helps to fill the void of pulling and manipulating real time data in the Hadoop space. So we have a set of products that can connect to sources of streaming data and act on it in real time. So using these kinds of products you can make things like trending videos on YouTube or trending hashtags on Twitter at this point in time. Some popular products in this space are Apache Spark (using the Spark Streaming module) and Apache Storm. We will be covering the Apache Spark streaming module in our chapter on real time analytics.

- **Machine learning libraries**: In the last few years there has been tremendous work in the predictive analytics space. Predictive analytics involves usage of advanced machine learning libraries and it's no wonder that some of these libraries are now included with the clustering computing frameworks as well. So a popular machine learning library such as Spark ML ships along with Apache Spark and older libraries such as Apache Mahout are also supported on big data. This is a growing space with new libraries frequently entering the market every few days.

- **NoSQL**: There are times when we need frequent reads and updates of data even though big data is involved. Under these situations there are a lot of non-SQL products that can be readily used while analyzing your data and some of the popular ones that can be used are Cassandra and HBase both of which are open source.
- **Search**: Quite often big data is in the form of plain text. There are many use cases where you would like to index certain words in the text to make them easily searchable. For example, if you are putting all the newspapers of a particular branch published for the last few years in HDFS in PDF format, you might want a proper index to be made over these documents so that they are readily searchable. There are products in the market that were previously used extensively for building search engines and they are now integratable with big data as well. One of the popular and open source options is SOLR and it can be easily established on top of big data to make the content easily searchable.

The categories of products we have just depicted previously is not extensive. We have not covered messaging solutions and there are many other products too apart from this. For checking on extensive lists refer to a book that specifically covers Hadoop in detail: for example, the *Hadoop Definitive Guide*.

We have covered the main categories of products, but let's now cover some of the important products themselves that are built on top of the big data stack:

Product	Description
HDFS	HDFS is a distributed filesystem that provides high-performance access to data across Hadoop clusters
Spark	The Spark cluster computing framework is used for various purposes such as analytics, stream processing, machine learning analytics, and so on, as shown in the preceding diagram.
Impala	Real-time data analytics is where you can fire queries in real time using this on big data; this is used by data scientists and business analysts.
MapReduce	MapReduce is a programming model and an associated implementation for processing and generating large datasets with a parallel, distributed algorithm on a cluster.
Sqoop	This helps to pull data from structured databases such as Oracle and push the data into Hadoop or HDFS
Oozie	This is a job scheduler for scheduling Hadoop jobs
Flume	This is a tool to pull large amount of streaming data into Hadoop/HSFS
Kafka	Kafka is a real-time stream processing engine which provides very high throughput and low latency.
Yarn	This is the resource manager in Hadoop 2

Distributed computing on Hadoop

Suppose you put plenty of data on a disk and read it. Reading this entire data takes, for example, 24 hours. Now, suppose you distribute this data on 24 different machines of the same type and run the read program at the same time on all the machines. You might be able to parallelly read the entire data in an hour (an assumption just for the purpose of this example). This is what parallel computing is all about though. It helps in processing large volumes of data parallelly on multiple machines called nodes and then combining the results to build a cumulated output. Disk input/output is so slow that we cannot rely on a single program running on one
machine to do all this for us.

There is an added advantage of data storage across multiple machines, which is failover and replication support of data.

The bare bones of Hadoop are the base modules that are shipped with its default download option. Hadoop consists of three main modules:

- **Hadoop core**: This is the main layer that has the code for the failover, data replication, data storage, and so on.

    HDFS: The **Hadoop Distributed File System** (**HDFS**) is the primary storage system used by Hadoop applications. HDFS is a distributed filesystem that provides high-performance access to data across Hadoop clusters.

- **MapReduce**: This is the data analysis framework that runs parallely on top of data stored in HDFS.

As you saw in the options above if you install the base Hadoop package you will get the core Hadoop library, the HDFS file system, and the MapReduce framework by default, but this is not extensive and the current use cases demand much more then the bare minimum products provided by the Hadoop default installation. It is due to this reason that a whole set of products have originated on top of this big data stack be, it the streaming products such as Storm or messaging products such as Kafka or search products such as SOLR.

HDFS concepts

HDFS is Hadoop's implementation of a distributed filesystem. The way it is built, it can handle large amount of data. It can scale to the extent where the other types of distributed filesystems, for example, NFS cannot scale to. It runs on plain commodity servers and any number of servers can be used.

HDFS is a write once, read several times type of filesystem. Also, you can append to a file, but you cannot update a file. So if you need to make an update, you need to create a new file with a different version. If you need frequent updates and the amount of data is small, then you should use other software such as RDBMS or HBASE.

Design and architecture of HDFS

These are some of the features of HDFS:

- **Open source**: HDFS is a completely open source distributed filesystem and is a very active open source project.
- **Immense scalability for the amount of data**: You can store petabytes of data in it without any problem.
- **Failover support**: Any file that is put in HDFS is broken into chunks (called blocks) and these blocks are distributed across different machines of the cluster. Apart from the distribution of this file data, the data is also replicated across the different machines depending upon the replication level. Thereby, in case any machine goes down; the data is not lost and is served from the other machine.
- **Fault tolerance**: This refers to the capability of a system to work in unfavorable conditions. HDFS handles faults by keeping replicated copies of data. So due to a fault, if one set of data in a machine gets corrupted then the data can always be pulled from some other replicated copy. The replica of the data is created on different machines, so even if the entire machine goes down, still is no problem as replicated data can always be pulled from some other machine that has the copy of it.
- **Data locality**: The way HDFS is designed, it allows the main data processing programs to run closer to the data where it resides and hence they are faster as less network transfer is involved.

Main components of HDFS

There are two main daemons that make up HDFS. They are depicted in the following diagram:

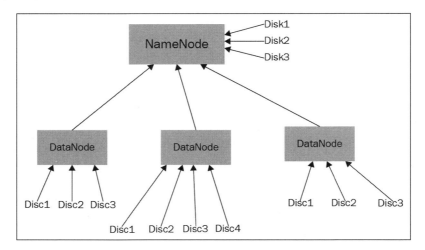

As you can see in the preceding diagram, the main components are:

- **NameNode**: This is the main program (master) of HDFS. A file in HDFS is broken in to chunks or blocks and is distributed and replicated across the different machines in the Hadoop cluster. It is the responsibility of the **NameNode** to figure out which blocks go where and where the replicated blocks land up. It is also responsible for clubbing the data of the file when the full file is asked for by the client. It maintains the full metadata for the file.

- **DataNodes**: These are the slave processes running on the other machines (other than the **NameNode** machine). They store the data and provide the data when the **NameNode** asks for it.

The most important advantage of this master/slave architecture of HDFS is failover support. Thereby, if any **DataNode** or slave machine is down, the **NameNode** figures this out using a heartbeat signal and it would then refer to another **DataNode** that has the replicated copy of that data. Before Hadoop 2, the **NameNode** was the single point of failure but after Hadoop 2, NameNodes have a better failover support. So you can run two NameNodes alongside one another so that if one **NameNode** fails, the other **NameNode** can quickly take over the control.

HDFS simple commands

Most of the commands on HDFS are for storing, retrieving, or discarding data on it. If you are used to working on Linux, then using HDFS shell commands is simple, as almost all the commands are a replica of the Linux commands with similar functions. Though the HDFS commands can be executed by the browser as well as using Java programs, for the purpose of this book, we will be only discussing the shell commands of HDFS, as shown in the following table:

Command	What it does
mkdir	This helps you to make a directory in HDFS: `hdfs dfs -mkdir /usr/etl` You always start the command with `hdfs dfs` and then the actual command, which is exactly similar to the Linux command. In this case, this command makes a directory `etl` inside the `/usr` directory in `hdfs`.
put	This helps you to copy a file from a local filesystem to `hdfs`: `hdfs dfs -put dataload1.txt /usr/etl` This copies a file `dataload1.txt` to `/usr/etl` directory inside `hdfs`
ls	This helps you to list out all files inside a directory: `hdfs dfs -ls /usr/etl` (lists out files inside `/usr/etl`)
rm	This helps you to remove a file: `hdfs dfs -rm /usr/etl/dataload.txt` (deletes `dataload.txt` inside `/usr/etl`)
du -h	This helps you to check the file size: `hdfs dfs -du -h /usr/etc/dataload.txt`
chmod	This helps you to change the permissions on all: `hdfs dfs -chmod 700 /usr/etl/dataload.txt` This only gives the owner of the file complete permissions; rest of the users won't have any permissions on the file.
cat	This helps you to read the contents of a file: `hdfs dfs -cat /usr/etl/dataload.txt`
head	This helps you to read the top content (few lines from top) of a file: `hdfs dfs -head /usr/etl/dataload.txt` Similarly, we have the `tail` command to read a few lines from the bottom of a file.
mv	This helps you to move a file across different directories: `hdfs dfs -mv /usr/etl/dataload.txt /usr/input/newdataload.txt`

Apache Spark

Apache Spark is the younger brother to the MapReduce framework. It's a cluster computing framework that is getting much more attraction now in comparison to MapReduce. It can run on a cluster of thousands of machines and distribute computations on the massive datasets across these machines and combine the results.

There are few main reasons why Spark has become more popular than MapReduce:

- It is way faster than MapReduce because of its approach of handling a lot of stuff in memory. So on the individual nodes of machines, it is able to do a lot of work in memory, but MapReduce on the other hand has to touch the hard disk many times to get a computation done and the hard disk read/write is slow, so MapReduce is much slower.
- Spark has an extremely simple API and hence it can be learned very fast. The best documentation is the Apache page itself, which can be accessed at spark.apache.org. Running algorithms such as machine learning algorithms on MapReduce can be complex but the same can be very simple to implement in Apache Spark.
- It has a plethora of sub-projects that can be used for various other operations.

Concepts

The main concept to understand Spark is the concept of RDDs or Resilient Distributed Dataset.

So what exactly is an RDD?

A **resilient distributed dataset** (**RDD**) is an immutable collection of objects. These objects are distributed across the different machines available in a cluster. To a Java developer, an RDD is nothing but just like another variable that they can use in their program, similar to an ArrayList. They can directly use it or call some actions on it, for example, count() to figure out the number of elements in it. Behind the job, it sparks tasks that get propagated to the different machines in the cluster and bring back the computed results in a single object as shown in the following example:

```
JavaRDD<String> rows = sc.textFile("univ_rankings.csv");
System.out.println("Total no. of rows --->"+ rowRdd.count());
```

The preceding code is simple yet it depicts the two powerful concepts of Apache Spark. The first statement shows a Spark RDD object and the second statement shows a simple action. Both of them are explained as follows:

- `JavaRDD<String>`: This is a simple RDD with the name `rows`. As shown in the `generics` parameter, it is of type `string`. So it shows that this immutable collection is filled with string objects. So, if Spark, in this case, is sitting on 10 machines, then this list of strings or RDD will be distributed across the 10 machines. But to the Java developer, this object is just available as another variable and if they need to find the number of elements or rows in it, they just need to invoke an action on it.

- `rows.count()`: This is the action that is performed on the RDD and it computes the total elements in the RDD. Behind the scene, this method would run on the different machines of the cluster parallelly and would club the computed result on each parallel node and bring back the result to the end user.

RDD can be filled with any kind of object, for example, Java or Scala objects.

Next we will cover the types of operations that can be run on RDDs. RDDs support two type of operations and they are transformations and actions. We will be covering both in the next sections.

Transformations

These are used to transform an RDD into just another RDD. This new RDD can later be used in different operations. Let's try to understand this using an example as shown here:

```
JavaRDD<String> lines = sc.textFile("error.log");
```

As shown in the preceding code, we are pulling all the lines from a `log` file called `error.log` into a `JavaRDD` of strings.

Now, suppose we need to only filter out and use the data rows with the word `error` in it. To do that, we would use a transformation and filter out the content from the `lines` RDD, as shown next:

```
JavaRDD<String> filtered = rowRdd.filter(s -> s.contains("error"));
System.out.println("Total no. of rows --->"+ filtered.count());
```

As you can see in the preceding code, we filtered the RDD based on whether the word `error` is present in its element or not and the new RDD `filtered` only contains the elements or objects that have the word `error` in it. So, transformation on one RDD produces another RDD only.

Actions

The user can take some actions on the RDD. For example, if they want to know the total number of elements in the RDD, they can invoke an action `count()` on it. It's very important to understand that until transformation, everything that happens on an RDD is in lazy mode only; that is, to say that the underlying data remains untouched until that point. It's only when we invoke an action on an RDD that the underlying data gets touched and an operation is performed on it. This is a design-specific approach followed in Spark and this is what makes it so efficient. We actually need the data only when we execute some action on it. What if the user filtered the `error` log for errors but never uses it? Then storing this data in memory is a waste, so thereby only when some action such as `count()` is invoked will the actual data underneath be touched.

Here are few common questions:

- When RDD is created, can it be reused again and again?

  An RDD on which no action has been performed but only transformations are performed can be directly reused again and again. As until that point no underlying data is touched in actuality. However, if an action has been performed on an RDD, then this RDD object is utilized and discarded as soon as it is used. As soon as an action is invoked on an RDD the underlying transformations are then executed or in other words the actual computation then starts and a result is returned. So an action basically helps in the return of a value.

- What if I want to re-use the same RDD even after running some action on it?

  If you want to reuse the RDD across actions, then you need to persist it or, in other words, cache it and re-use it across operations. Caching an RDD is simple. Just invoke an API call persist and specify the type of persistence. For example, in memory or on disk, and so on. Thereby, the RDD, if small, can be stored in the memory of the individual parallel machines or it could be written to a disk if it is too big to fit into memory.

  An RDD that is stored or cached in this way, as mentioned earlier, is reusable only within that session of Spark Context. That is, to say if your program ends the usage ends and all the temp disk files of the storage of RDD are deleted.

- So what would you do if you need an RDD again and again in multiple programs going forward in different `SparkContext` sessions?

  In this case, you need to persist and store the RDD in an external storage (such as a file or database) and reuse it. In the case of big data applications, we can store the RDD in HDFS filesystem or we can store it in a database such as HBase and reuse it later when it is needed again.

  In real-world applications, you would almost always persist an RDD in memory and reuse it again and again to expedite the different computations you are working on.

- What does a general Spark program look like?

  Spark is used in massive ETL (extract, transform, and load), predictive analytics, or reporting applications.

  Usually the program would do the following:
  1. Load some data into the RDD.
  2. Do some transformation on it to make the data compatible to handle your operations.
  3. Cache the reusable data across sessions (by using persist).
  4. Do some actions on the data; the action can be ready-made or can be custom operations that you wrote in your programs.

Spark Java API

Since Spark is written in Scala, which inherently is written in Java, Java is the big brother on the Apache Spark stack and is fully supported on all its products. It has an extensive API on the Apache Spark, stack. On Apache Spark Scala is a popular language of choice but most enterprise projects within big corporations still heavily rely on Java. Thus, for existing java developers on these projects, using Apache Spark and its modules by their java APIs is relatively easy to pick up. Here are some of the Spark APIs that java developers can easily use while doing their big data work:

- Accessing the core RDD frameworks and its functions
- Accessing Spark SQL code
- Accessing Spark Streaming code
- Accessing the Spark GraphX library
- Accessing Spark MLlib algorithms

Apart from this, Java is very strong on the other big data products as well. To show how strong Java is on the overall big data scene, let's see some examples of big data products that readily support Java:

- **Working on HBase using Java**: HBase has a very strong java API and data can easily be manipulated on it using Java
- **Working on Hive using Java**: Hive is a batch storage product and working on it using Java is easy as it has a good Java API.
- Even HDFS supports a Java API for regular file handling operations on HDFS.

Spark samples using Java 8

All our samples in the book are written using Java 8 on Apache Spark 2.1. Java 8 is aptly suited for big data work mainly because of its support for lambda's, due to which the code is very concise. In the older versions of Java, the Apache Spark Java code was not concise but Java 8 has changed completely.

We will encourage the readers of this book to actively use the Java 8 API on Apache Spark as it not only produces concise code, but overall improves the readability and maintainability of the code. One of the main reasons why scala is heavily used on Apache Spark was mainly due to the concise and easy to use API. But with the usage of Java 8 on Apache Spark, this advantage of Scala is no longer applicable.

Loading data

Before we use Spark for data analysis, there is some boilerplate code that we always have to write for creating the `SparkConfig` and creating the `SparkContext`. Once these objects are created, we can load data from a directory in HDFS.

For all real-world applications, your data would either reside in HDFS or in databases such as Hive/HBase for big data.

Spark lets you load a file in various formats. Let's see an example to load a simple CSV file and count the number of rows in it.

We will first initialize a few parameters, namely, application name, master (whether Spark is locally running this or on a cluster), and the data filename as shown next:

```
private static String appName =LOAD_DATA_APPNAME";
private static String master =local";
private static String FILE_NAME =univ_rankings.txt";\
```

Next, we will create the `SparkContext` and Spark config object:

```
SparkConf conf =new
SparkConf().setAppName(appName).setMaster(master);
JavaSparkContext sc =new JavaSparkContext(conf);
```

Using the `SparkContext`, we will now load the data file:

```
JavaRDD<String> rowRdd = sc.textFile(FILE_NAME);
```

Data operations – cleansing and munging

This is the task on which the data analyst would be spending the maximum amount of time on. Most of the time, the data that you would be using for analytics will come from `log` files or will be generated from other data sources. The data won't be clean and some data entries might be missing or incorrect completely. Before any data analytic tasks can be run on the data, it has to be cleaned and prepared in good shape for the analytic algorithms to run on. We will be covering cleaning and munging in detail in the next chapter.

Analyzing data – count, projection, grouping, aggregation, and max/min

I assume that you already have Spark installed. If not, refer to the Spark documentation on the web for installing Spark on your machine. Let's now use some popular transformation and actions on Spark.

For the purpose of the following samples, we have used a small dataset of university rankings from `Kaggle.com`. It can be download from this link: https://www.kaggle.com/mylesoneill/world-university-rankings. It is a comma-separated dataset of university names followed by the country the university is located at. Some sample data rows are shown next:

```
Harvard University, United States of America

California Institute of Technology, United States of America

Massachusetts Institute of Technology, United States of America …
```

Common transformations on Spark RDDs

We will now cover a few common transformation operations that we frequently run on the RDDs of Apache Spark:

1. **Filter**: This applies a function to each entry of the RDD, for example:

   ```
   JavaRDD<String> rowRdd = sc.textFile(FILE_NAME);
   System.out.println(rowRdd.count());
   ```

   As shown in the preceding code, we loaded the data file using Spark context. Now, using the `filter` function we will filter out the rows that contain the word `Santa Barbara` as shown next:

   ```
   JavaRDD<String> filteredRows = rowRdd.filter(s ->
   s.contains("Santa Barbara"));
   System.out.println(filteredRows.count());
   ```

2. **Map**: This transformation applies a function to each entry of an RDD.

3. In the RDD we read earlier we will find the length of each row of data using the `map` function as shown next:

   ```
   JavaRDD<Integer> rowlengths = rowRdd.map(s -> s.length());
   ```

   After reading the length of each row in the RDD, we can now collect the data of the RDD and print its content:

   ```
     List<Integer> rows = rowlengths.collect();
   for(Integer row : rows){
             System.out.println(row);
   }
   ```

4. **FlatMap**: This is similar to map, except, in this case, the function applied to each row of RDDs will return a list or sequence of values instead of just one, as in case of the preceding `map`. As an example, let's create a sample RDD of strings using the `parallelize` function (this is a handy function for quick testing by creating dummy RDDs):

   ```
   JavaRDD<String> rddX = sc.parallelize(
   Arrays.asList("big data","analytics","using java"));
   ```

   On this RDD, let's split the strings by the spaces between them:

   ```
   JavaRDD<String[]> rddY = rddX.map(e -> e.split(" "));
   ```

   Finally, `flatMap` will connect all these words together into a Single List of object as follows:

   ```
   {"big","data","analytics","using","java"}

   JavaRDD<String> rddY2 = rddX.flatMap(e ->
   Arrays.asList(e.split(" ")).iterator());
   ```

We can now collect and print this `rddY2` in a similar way as shown here for other RDDs.

5. Other common transformations on RDDs are as follows:

Other transformation	Description
Union	This is a union of two RDDs to create a single one. The new RDD is a union set of both the other RDDs that are combined.
Distinct	This creates an RDD of only distinct elements.
Map paritions	This is similar to a map as shown earlier, but runs separately on each partition block of the RDD.

Actions on RDDs

As mentioned earlier, the actual work on the data starts when an action is invoked. Until that time, all the transformations are tracked on the driver program and sent to the data nodes as a set of tasks.

We will now cover a few common **actions** that we frequently run on the RDDs of Apache Spark:

- `count`: This is used to count the number of elements of an RDD.

    For example, the `rowRdd.count()method` would count the rows in row RDD.

- `collect`: This brings back all the data from different nodes into an array on the driver program (It can cause memory leaks on the driver if the driver is low on memory.). This is good for quick testing on small RDDs:

    ```
    JavaRDD<String> rddX = sc.parallelize(
    Arrays.asList("big data","analytics","using java"));
            List<String> strs = rddX.collect();
    ```

    This would print the following three strings:

    ```
    'Big data
     Analytics
     Using java'
    ```

- `reduce`: This action takes in two parameters and returns one. It is used in aggregating the data elements of an RDD. As an example, let's create a sample RDD using the `parallelize` function:

    ```
    JavaRDD<String> rddX2 =
    sc.parallelize(Arrays.asList("1","2","3"));
    ```

After creating the RDD `rddX2`, we can sum up all its integer elements by invoking the `reduce` function on this RDD:

```
String sumResult = rddX2.reduce((String x, String y) ->
{
return»»+(Integer.parseInt(x)+ Integer.parseInt(y));
});
```

Finally, we can print the sum of RDD elements:

```
System.out.println("sumResult ==>"+sumResult);
```

- `foreach`: Just as the `foreach` loop of Java works in a collection, similarly this action causes each element of the RDD to be accessed:

  ```
  JavaRDD<String> rddX3 = sc.parallelize(
              Arrays.asList("element-1","element-2","element-3"));
          rddX3.foreach(f -> System.out.println(f));
  ```

This will print the output as follows:

```
element-1
element-2
element-3
```

Paired RDDs

As HashMap is a key-value pair collection, similarly, paired RDDs are key-value pair collections except that the collection is a distributed collection. Spark treats these paired RDDs specially and provides special operations on them as shown next.

An example of a paired RDD:

Let's create a sample key-value paired RDD using the `parallelize` function:

```
JavaRDD<String> rddX = sc.parallelize(
            Arrays.asList("videoName1,5","videoName2,6",
"videoName3,2","videoName1,6"));
```

Now, using the `mapToPair` function, extract the keys and values from the data rows and return them as an object of a key-value pair or simple a `Tuple2`:

```
JavaPairRDD<String, Integer> videoCountPairRdd = rddX.
mapToPair((String s)->{
            String[] arr = s.split(",");
return new Tuple2<String, Integer>(arr[0],
Integer.parseInt(arr[1]));
});
```

Now, collect and print these rules:

```
List<Tuple2<String,Integer>> testResults =
videoCountPairRdd.collect();
for(Tuple2<String, Integer> tuple2 : testResults){
    System.out.println(tuple2._1);
}
```

This will print the output as follows:

```
videoName2
videoName3
videoName1
```

Transformations on paired RDDs

Just as we can run transformations on plain RDDs we can also run transformations on top of paired RDDs too. Some of the transformations that we can run on paired RDDs are explained as follows:

- `reduceByKey`: This is a transformation operation on a key-value paired RDD. This operation involves shuffling across different data partitions and creates a new RDD. The parameter to this operation is a cumulative function, which is applied on the elements and an aggregation is done on those elements to produce a cumulative result.

    In the preceding RDD, we have data elements for video name and hit counts of the videos as shown in the following table:

Video name	Hit counts.
videoName1	5
videoName2	6
videoName3	2
videoName1	6

    We will now try to run `reduceByKey` on the paired RDD to find the net hit counts of all the videos as shown earlier.

    We will be loading the data into an RDD in the same way as shown earlier. Once the data is loaded, we can do a `reduceByKey` to sum up the hit counts on the different videos:

    ```
    JavaPairRDD<String, Integer> sumPairRdd =
    videoCountPairRdd.reduceByKey((x,y) -> x + y);
    ```

After the transformation, we can collect the results and print them as shown next:

```
List<Tuple2<String,Integer>> testResults = sumPairRdd.collect();
for(Tuple2<String, Integer> tuple2 : testResults){
    System.out.println("Title : "+ tuple2._1 +
", Hit Count : "+ tuple2._2);
}
```

The results should be printed as follows:

```
Title : videoName2, Hit Count : 6
Title : videoName3, Hit Count : 2
Title : videoName1, Hit Count : 11
```

- groupByKey: This is another important transformation on a paired RDD. Sometimes, you want to club all the data for a particular key into one iterable unit so that you can later go through it for some specific work. groupByKey does this for you, as shown next:

```
JavaPairRDD<String, Iterable<Integer>> grpPairRdd =
videoCountPairRdd.groupByKey();
```

After invoking groupByKey on videoCountPairRdd, we can collect and print the result of this RDD:

```
List<Tuple2<String,Iterable<Integer>>> testResults = grpPairRdd.collect();
for(Tuple2<String, Iterable<Integer>> tuple2 : testResults){
System.out.println("Title : "+ tuple2._1 );
        Iterator<Integer> it = tuple2._2.iterator();
int i =1;
while(it.hasNext()){
            System.out.println("value "+ i +" : "+ it.next());
            i++;
}
}
```

And the results should be printed as follows:

```
Title : videoName2
value 1 : 6
Title : videoName3
value 1 : 2
Title : videoName1
value 1 : 5
value 2 : 6
```

As you can see, the contents of the `videoName1` key were grouped together and both the counts 5 and 6 were printed together.

Saving data

The contents of an RDD can be stored in external storage. The RDD can later be rebuilt from this external storage too. There are a few handy methods for pushing the contents of an RDD into external storage, which are:

- `saveAsTextFile`(path): This writes the elements of the dataset as a `text` file to an external directory in HDFS
- `saveAsSequenceFile`(path): This writes the elements of the dataset as a Hadoop SequenceFile in a given path in the local filesystem—HDFS or any other Hadoop-supported filesystem

Collecting and printing results

We have already seen in multiple examples earlier that by invoking `collect()` on an RDD, we can cause the RDD to collect data from different machines on the cluster and bring the data to the driver. Later on, we can print this data too.

When you fire a collect on an RDD at that instant the data from the distributed nodes is pulled and brought into the main node or driver nodes memory. Once the data is available, you can iterate over it and print it on the screen. As the entire data is brought in memory this method is not suitable for pulling a heavy amount of data as that data might not fit in the driver memory and an out of memory error might be thrown. If the amount of data is large and you want to peek into the elements of that data then you can save your RDD in external storage in Parquet or text format and later analyze it using analytic tools like Impala or Spark SQL. There is also another method called `take` that you can invoke on the Spark RDD. This method allows you to pull a subset of elements from the first element of the arrays. Thereby `take` method can be used when you need to view just a few lines from the RDD to check if your computations are good or not.

Executing Spark programs on Hadoop

Apache Spark comes with a script `spark-submit` in its `bin` directory. Using this script, you can submit your program as a job to the cluster manager (such as Yarn) of Spark and it would run this program. These are the typical steps in running a Spark program:

1. Create a `jar` file of your Spark Java code.

2. Next, run the `spark-submit` job by giving the location of the `jar` file and the main class in it. An example of the command is shown next:

   ```
   ./bin/spark-submit --class <main-class> --master <master-url><application-jar>
   ```

Some of the commonly used options of `spark-submit` are shown in the following table:

spark-submit options	What it does
`--class`	Your Java class that is the main entry point for the spark code execution.
`--master`	Master URL for the cluster
`application-jar`	Jar file containing your Apache spark code

 For additional `spark-submit` options, please refer to the Spark programming guide on the web. It has extensive information on it.

Apache Spark sub-projects

Apache Spark has now become a complete ecosystem of many sub-projects. For different operations on it, we have different products as shown next:

Spark sub-module	What it does
Core Spark	This is the foundation framework for all the other modules. It has the implementation for Spark computing engine, that is, RDD, executors, storage, and so on.
Spark SQL	Spark SQL is a Spark module for structured data processing. Using this you can fire SQL queries on your distributed datasets. It's very easy to use.
Spark Streaming	This module helps in processing live data streams, whether they are coming from products such as Kafka, Flume, or Twitter.
GraphX	Helps in building components for Spark parallel graph computations.
MLlib	This is a machine learning library that is built on the top of the Spark core and hence the algorithms are parallelly distributable across the massive datasets.

Spark machine learning modules

Spark MLlib is Spark's implementation of the machine learning algorithms based on the RDD format. It consists of the algorithms that can be easily run across a cluster of computer machines parallelly. Hence, it is much faster and scalable than single node machine learning libraries such as `scikit-learn`. This module allows you to run machine learning algorithms on top of RDDs. The API is not very user friendly and sometimes it is difficult to use.

Recently Spark has come up with the new Spark ML package, which essentially builds on top of the Spark dataset API. As such, it inherits all the good features of the datasets that are massive scalability and extreme ease of usage. If anybody has used the very popular Python scikit library for machine learning, they would realize that the API of the new Spark ML is quite similar to Python scikit. From the Spark documentation, Spark ML is the recommended way for doing machine learning tasks now and the old Spark MLlib RDD based API would get deprecated in some time.

Spark ML being based on datasets allows us to use Spark SQL along with it. Feature extraction and feature manipulation tasks become very easy as a lot can now be handled using Spark SQL only, especially the data manipulation work using plain SQL queries. Apart from this, Spark ML ships with an advanced feature called Pipeline. Plain data is usually in an extremely raw format and this data usually goes through a cycle or workflow where it gets cleaned, mutated, and transformed before it is used for consumption and training of machine learning models. This entire workflow of data and its stages is very well encapsulated in the new feature called as Pipeline in the Spark ML library. So you can work on the different workflows whether for feature extraction, feature transformation or converting features to mathematical vector format and gel together all this code using the pipeline API of Spark ML. This helps us in maintaining large code bases of machine learning stacks, so if later on you want to switch some piece of code (for example, for feature extraction), you can separately change it and then hook it into the pipeline and this would work cleanly without changing or impacting any other area of code.

MLlib Java API

The MLlib module is completely supported in Java and it is quite easy to use.

Other machine learning libraries

There are many machine learning libraries currently out there. Some of the popular ones are `scikit-learn`, `pybrain`, and so on. But as I mentioned earlier, these are single node libraries that are built to run on one machine but the algorithms are not optimized to run parallelly across a stack of machines and then club the results.

> How do you use these libraries on big data in case there is a particular algorithm implementation that you want to use from these libraries?
>
> On all the parallel nodes that are running your Spark tasks, make sure the particular installation of the specific library is present. Also any jars or executables that are required to run the algorithm must be available in the path of the `spark-submit` job to run this.

Mahout – a popular Java ML library

Apache Mahout is also a popular library and is open source from the Apache stack. It contains scalable machine learning algorithms. Some of the algorithms can be used for tasks such as:

- Recommendations
- Classfications
- Clustering

Some important features of Mahout are as follows:

- Its algorithms run on Hadoop so they work well in a distributed environment
- It has MapReduce implementations of several algorithms

Deeplearning4j – a deep learning library

This library is fully built on Java and is a deep learning library. We will cover this library in our chapter on deep learning.

Compressing data

Big data is distributed data that is spread across many different machines. For various operations running on data, data transfer across machines is a given. These are the formats supported on Hadoop for input compression: gzip, bzip, snappy, and so on. While we won't go into detail for the compression piece, it must be understood that when you actually work on big data analytics tasks, compressing your data will be always beneficial, providing few main advantages as follows:

- If the data is compressed, the data transfer bandwidth needed is less and as such the data would transfer fast.
- Also, the amount of storage needed for compressed data is much less.

- Hadoop ships with a set of compression formats that support easy distributability across a cluster of machines. So even if the compressed files are chuncked and distributed across a cluster of machines, you would be able to run your programs on them without loosing any information or important data points.

Avro and Parquet

Spark helps in writing the data to Hadoop and in Hadoop input/output formats. Avro and Parquet are two popular Hadoop file formats that have specific advantages. For the purpose of our examples, other than the usual file formats of data, such as log and text format, the files can also be present in Avro or Parquet format.

So what is Avro and Parquet and what is special about them?

Avro is a row-based format and is also schema based. The schema for the structure of the data row is stored within the file; due to this, the schema can independently change and there won't be any impact on reading old files. Also, since it is in row-based format, the files can easily be split, based on rows and put on multiple machines and processed parallely. It has good failover support too.

Parquet is a columnar file format. Parquet is specifically suited for applications where for analytics you only need a subset of your columnar data and not all the columns. So for things such as summing up/aggregating specific column Parquet is best suited for such operations. Since Parquet helps in choosing only the columns that are needed, it reduces disk I/O tremendously and hence it reduces the time for running analytics on the data.

Summary

In this chapter, we covered what big data is all about and how we can analyze it. We showed the 3 Vs that constitute big data: volume, variety, and velocity. We also covered some ground on the big data stack, including Hadoop, HDFS, and Apache Spark. While learning Spark, we went through some examples of the Spark RDD API and also learned a few useful transformations and actions.

In the next chapter, we will get the first taste of running analytics on big data. For this, we will initially use Spark SQL, a very useful Spark module, to do simple yet powerful analysis of your data and later we will go on to build complex analytic tasks while learning market basket analysis.

2
First Steps in Data Analysis

Let's take the first steps towards data analysis now. Spark has a very useful module, Spark. Apache Spark has a prebuilt module called as Spark SQL and this module is used for structured data processing. Using this module, we can execute SQL queries on our underlying data. Spark lets you read data from various datasources whether text, CSV, or Parquet files on HDFS or also from hive tables or HBase tables. For simple data analysis tasks, whether you are exploring your datasets initially or trying to analyze and cut a report for your end users with simple stats this module is tremendously useful.

In this chapter, we will work on two datasets. The first dataset that we will analyze is a simple dataset and the next one is a more complex real-world dataset from an e-commerce store.

In this chapter, we will cover the following topics:

- Basic statistical analytic approaches using Spark SQL
- Building association rules using the Apriori algorithm
- Advantages and disadvantages of using the Apriori algorithm
- Building association rules using a faster and more efficient FP-Growth algorithm

Datasets

Before we get our hands wet in the world of complex analytics, we will take small baby steps and learn some basic statistical analysis first. This would help us get familiar with the approach that we will be using on big data for other solutions as well. For our analysis initially we will take a simple cars JSON dataset that has details about a few cars from different countries. We will analyze it using Spark SQL and see how easy it is to query and analyze datasets using Spark SQL. Spark SQL is handy to use for basic analytics purposes and is nicely suited on big data. It can be run on massive datasets and data can reside in HDFS.

First Steps in Data Analysis

To start with a simple case study we are using a `cars` dataset. This dataset can be obtained from `http://www.carqueryapi.com/`. It can be obtained from link `http://www.carqueryapi.com/api/0.3/?callback=?&cmd=getMakes`. This datasets contains data about cars in different countries. It is in JSON format. It is not a very big dataset from the perspective of big data but for our learning purposes to start with a simple analytics case study it suits our requirements well. This dataset has four important attributes shown as follows:

Attribute name	Attribute description
`make_id`	The type of car, for example Acura, Mercedes
`make_display`	Name of the mode of the car
`make_is_common`	Check if the makel is a common model (marked as `1` if it is a common model else `0`)
`make_country`	Country where the car is made

> We are using this data only for learning purposes. The cars dataset can be replaced by any other dataset too for our learning purposes here. Hence we are not bothered about the accuracy of this data and we are not using it other than for our simple learning case study here.

Also here is a sample of some of the data in the dataset:

Sample row of dataset	Description
```{ "make-id":"acura", "make_display": "Acura", "make_is_common":"1", "make_country" : "USA" }```	Here, the `make_id` or type of car is acura and it is made in the country USA.
```{ "make-id":"alfa romeo", "make_display": "AlfaRomeo", "make_is_common":"1", "make_country" : "Italy" }```	Here, the car is of type `AlfaRomeo` and it is made in `Italy`.

[30]

Data cleaning and munging

The major amount of time spent by a developer while performing a data analysis task is spent in data cleaning or producing data in a particular format. Most of the time, while performing analysis of some `log` file data or getting files from some other system, there will definitely be some data cleaning involved. Data cleaning can be in many forms whether it involves discarding a certain kind of data or converting some bad data into a different format. Also note that most of the machine learning algorithms involve running algorithms on a mathematical dataset, but most of the practical datasets won't always have mathematical data. Converting text data to mathematical form is another important task that many developers need to do themselves before they can apply the data analysis tasks on the data.

If there are problems in the data that we need to resolve before we use it, then this approach of fixing the data is called as data munging. One of the common data munging tasks is to fix up null values in data and these null values might represent either bad data or missing data. Bad or missing data is not good for our analysis as it can result in bad analytical results. These data issues need to be fixed before we can use our data in actual analysis. To learn the concepts of how we can fix our data before we use it in our analysis let's pick up the dataset that we are using in this chapter and fix the data before analyzing these datasets.

Most of your time as a developer performing the task of data analysis on big data will be spent on making the data good for training the models. The general tasks might include:

- **Filtering the unwanted data**: There are times when some of the data in your dataset might be corrupted or might be bad. If you can fix this data somehow, then you should, else you will have to discard it. Sometimes the data might be good but it might contain attributes that you don't need. In this case, you can discard these extra attributes. You can also use the Apache Spark's `filter` method to filter out the unwanted data.

- **Handling incomplete or missing data**: Not all data points might be present in the data. In such a situation, the developer needs to figure out which data point or default data point is needed when the data point is not available. Filling missing values is a very important task especially if you are using this data to analyze your dataset. We will look at some of the common strategies for handling missing data.

- **Discarding data**: If a lot of attributes in the data are missing, one easy approach is to discard this row of data. This is not a very fruitful approach especially if there are some attributes within this row that are meaningful, which we are using.

- **Fill some constant value**: You can fill in some constant generic value for missing attributes; for example, in your car, if you have entries as shown in the following table with empty `make_id` and empty `make_display`:

Dataset one sample row
{
"make_id ";"",
"make_display "; "",
"make_country" ;" JAPAN"
}

  If we discard these entries, it won't be a good approach. If we are asked to find the total number of cars from JAPAN in this dataset, then we will use the following code:

  ```
  make_country = 'JAPAN'.
  ```

  To counter this and use this data, we can fill in some constant value such as Unknown in this field. So the field will look like this:

  ```
  { "make_id ";"Unknown", "make_display "; "UnKnown", "make_country" ;" JAPAN" }
  ```

  As shown earlier, we have filled the UnKnown keyword wherever we saw empty data as in the case of `make_id` and `make_display`.

- **Populate with average value**: This might work in some cases. So if you have a missing value in some column, you can take all the values with good data in that column and find an average and later use this average value as a value on that item.

- **Nearest Neighbor approach**: This is one of my favorite approaches, and once we cover the KNN algorithm in this book we will cover this topic again. Basically, you find data points that are similar to the one with missing attributes in your dataset. You then replace the missing attributes with the attributes of the nearest data point that you found. So suppose you have your data from the dataset plotted on a scatter plot, as shown in the following screenshot:

The preceding screenshot shows some data points of a dataset plotted on the x and y axis on a scatter plot. Look at the datapoint as shown by the arrow with **Point A** as label. If this datapoint has some missing attributes, then we find the nearest data point to it which in this case is datapoint B as shown by the other arrow (which has **Point B** as a label). From this datapoint, we now pull the missing attributes. For this approach, we use the KNN algorithm or the K Nearest Neighbor algorithm to figure out the distance of one data point from another based on some attributes:

- **Converting data to a proper format**: Sometimes you might have to convert data from one format to another for your analytics task. For example, converting non-numeric numbers to numeric numbers or converting the date field to a proper format.

Basic analysis of data with Spark SQL

Spark SQL is a spark module for structured data processing. Almost all the developers know SQL. Spark SQL provides an SQL interface to your Spark data (RDDs). Using Spark SQL you can fire SQL queries or SQL-like queries on your big data set and fetch data in objects called dataframes.

A dataframe is like a relational database table. It has columns in it and we can apply functions to these columns such as `groupBy`, and so on. It is very easy to learn and use.

In the next section, we will cover a few examples on how we can use the dataframe and run regular analysis tasks.

Building SparkConf and context

This is just boilerplate code and is the entry point for the usage of our Spark SQL code. Every spark program will start with this boiler plate code for initialization. In this code we build the Spark configuration and then apply the configuration parameters (like application name and master location) and also build the SparkSession object. This SparkSession object is the main object using which you can fire SQL queries on your dataset.

```
SparkConf sconf = new sparkConf().setAppName(APP_NAME) .setMaster(APP_MASTER);
SparkSession spark = SparkSession.builder() .config(sconf) .getOrCreate();
```

Dataframe and datasets

Dataframe is a collection of distributed objects organized into named columns. It is similar to a table in a relational database and you can use Spark SQL to query it in a similar way. You can build dataframes from various datasources such as JSON files, CSV files, parquet files or directly from Hive tables, and so on.

A dataset is also a collection of distributed objects, but is essentially a hybrid of a **Resilient Distributed Dataset** (**RDD**) and a dataframe. An RDD or resilient distributed dataset is a distributed collection of objects, is similar to an array list in Java except that it is filled with objects that are distributed across multiple machines. Spark provides low level API to interact with this distributed object. Dataframe on the other hand is a higher level abstraction on top of RDDs and they are similar to relational database tables which store data in that format. SQL queries can be fired on top of dataframes. As we mentioned before a dataset object is a hybrid of dataframe and RDD and it supports firing SQL queries similar to dataframes and also applying RDD functions such as `map`, `filter`, and `flatMap`, similar to RDDs.

Load and parse data

Spark API is very extensive. We can load data out of the box in different formats and can clean/munge the data as we require and use it in our analysis tasks. The following code shows us ways of loading different datasets. Here we are loading data from a JSON file. This builds `Dataset<Row>` which is similar to a table in a relational database, it has a set of columns:

```
Dataset<Row> carsBaseDF = spark.read() .json("src/resources/data/cars.json");
            carsBaseDF.show();
```

Now we will register this dataframe as a temporary view. Just registering it as a `temp` table in `SparkContext` means we can fire queries on it just as you execute queries on an RDBMS table. That's as simple as it gets. To use this dataset row as a relational database table and fire queries on it, just use the `createOrReplaceTempView` method shown as follows:

```
carsBaseDF.createOrReplaceTempView("cars");
```

Now this data is available as a table `cars` just like a relational database table and you can fire any SQL queries on it such as `select * from cars` to pull all the rows.

Analyzing data – the Spark-SQL way

Let's now dive into a few examples. You can find more examples in the accompanying code in the GitHub repository too. For brevity, I am not showing the boilerplate code for `SparkContext` again and again. I will be just referring to `SparkSession` object as `spark`:

- **Simply select and print data**: Here we will just execute a query on the `cars` table and would print a sample result from the entire dataset of results. It's exactly similar to firing a `select` query on a relational database table:

```
Dataset<Row> netDF = spark.sql("select * from cars");
            netDF.show ();
```

First Steps in Data Analysis

The result will be printed as follows:

```
+------------+------------+----------+--------------+
|make_country| make_display|   make_id|make_is_common|
+------------+------------+----------+--------------+
|       Italy|      Abarth|    abarth|             0|
|          UK|          AC|        ac|             0|
|         USA|       Acura|     acura|             1|
|       Italy|   AlfaRomeo|alfa-romeo|             1|
|          UK|      Allard|    allard|             0|
+------------+------------+----------+--------------+
only showing top 5 rows
```

- **Filtering on data**: Here I will show two simple ways for filtering the data. First we will select a single column and print results from the top few rows. For this we will use the spark session and fire a SQL query on the cars table. We will be selecting only the two columns make_country and make_display from the cars table shown next. Also, for printing the first few rows, we will use a handy spark method show(), which will print the first few rows of the result set:

```
Dataset<Row> singleColDF =
spark.sql("select make_country,make_display fromcars") ;
singleColDF.show();
```

The output is as follows:

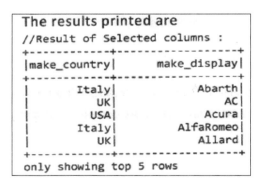

```
The results printed are
//Result of Selected columns :
+------------+------------+
|make_country|make_display|
+------------+------------+
|       Italy|      Abarth|
|          UK|          AC|
|         USA|       Acura|
|       Italy|   AlfaRomeo|
|          UK|      Allard|
+------------+------------+
only showing top 5 rows
```

- **Total count**: Here we will find the total rows in our dataset. For this we will use the count method on the dataset. The count method when executed on the dataset returns the total number of rows in the dataset.

```
System.out.println("Total Rows in Data --) " + netDF.count();
```

The output is as follows:

```
Total Rows in dataset :155
```

- **Selective data**: Let's fetch some data based on some criteria:
    - **Fetch the cars made in Italy only**: We will fire a query on our `car` view with a `where` clause specifying the `make_country` as `'Italy'`:

        ```
        Dataset<Row> italyCarsDF =
        spark.sql("select * from cars where make_country 'Italy'"};
                    italyCarsDF.show(}; //show the full content
        ```

        The result will be printed as follows:

        ```
        +------------+------------+------------+--------------+
        |make_country|make_display|    make_id |make_is_common|
        +------------+------------+------------+--------------+
        |       Italy|      Abarth|      abarth|             0|
        |       Italy|   AlfaRomeo|  alfa-romeo|             1|
        |       Italy| Autobianchi| autobianchi|             0|
        |       Italy|   Bizzarrini|  bizzarrini|             0|
        |       Italy|     Bugatti|     bugatti|             1|
        |       Italy|   De Tomaso|   de-tomaso|             0|
        |       Italy|     Ferrari|     ferrari|             1|
        |       Italy|        Fiat|        fiat|             1|
        |...
        +------------+------------+------------+--------------+
        ```

    - **Fetch the count of cars from Italy**: We will just use the `count` method on the dataset we received in the previous call where we fetched the rows that belonged only to country `'Italy'`:

        ```
        System.out.println("Data on Italy Cars");
        System.out.println ("Number of cars from Italy in this data set --> " +
        italyCarsDF.count ();
        ```

        This will print the following:

        ```
        Number of cars from Italy in this dataset --> 17
        ```

- **Collect all data and print it**: Now discard the `show()` function as it is just a handy function for testing and instead of that let's use a function that we will use to get the data after firing the queries.

  ```
  List<Row> italyRows = italyCarsDF.collectAsList();
  for (Row italyRow : italyRows) {
  System.out.println("Car type -> " + italyRow.getString(1);
  }
  ```

  This will print out all the types of cars that are made in Italy as shown (we are only showing the first few cars here)

  ```
  Car type -> Abarth
  Car type -> AlfaRomeo
  Car type -> Autobianchi
  Car type -> Bizzarinni
  Car type -> Bugatti
  ```

- **Total count of cars from Japan in the dataset**: We selected records that belong to Italy. Let's find the total count of cars from `Japan` in the dataset. This time we will just pull the count and not the total data for Japanese cars:

  ```
  Dataset<Row> jpnCarsDF =
  spark.sql("select count(*) from cars where make_country = 'Japan'");
  List<Row> jpnRows = jpnCarsDF.collectAsList();
          System.out.println("Japan car dataset ------> " +
  jpnRows.get(0).getLong(0);
  ```

  As shown, we build a dataframe by searching only for Japanese cars, and next we print the count of these rows. The result is as follows:

  ```
  Japan car dataset ------> 15
  ```

- **Distinct countries and their count**: Just like we use the `distinct` clause in SQL we can use the `distinct` clause in this big data Spark SQL query. If the result is small, as in this case, we can do a `collect()` and bring the data result in the memory of the driver program and print it there.

Using the following code, we will print the distinct countries in this dataset of cars:

```
Dataset<Row> distinctCntryDF = spark.sql("select distinct make_country from
Cars");

List<Row> distinctCtry = distinctCntryDF.collectAsList();
System.out.println("Printing Distinct Countries below");

for (Row drow : distinctCtry) {
System.out.println(drow.get(0).toString();
System. out. println( "Total Distinct Countries ; " +
distinctCtry.length);
}
```

And the result is printed as follows:

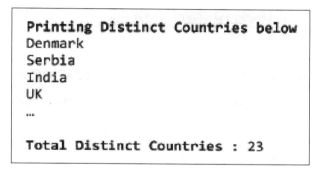

- **Group by country and find count**: Now, let's try to find the number of cars from each country and sort it in descending order. As most Java developers have used SQL before, this is a simple group by clause along with an order by for ordering by count in descending order as shown:

    ```
    Dataset<Row> grpByCntryDF = spark.sql("select
     make_country,count(*) cnt from Cars order by cnt
    desc");
    ```

    As seen we fired a simple group by query and counted the number of countries in the dataset and finally sorted by the count in descending order. We will now print the first few rows of this dataset:

    ```
    grpByCntryDF.show()
    ```

The result should be printed as follows:

```
UK , 39
USA , 29
Italy , 17
Germany , 16
Japan , 15
France , 8
South Korea , 5
Netherlands , 3
China , 3
Sweden , 3
Russia , 3
India , 2
Czech Republic , 2
...
```

There is a saying that a picture says a thousand words. Let's plot this in a bar chart and see how easy it is to visualize this data by country:

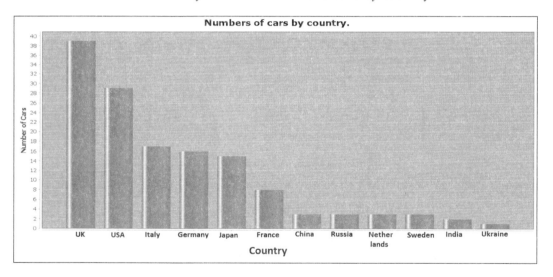

As you can see from the graph, it is very easy to figure out that the UK has the maximum number of cars in this dataset followed by USA.

We will be covering graphs for visualization in detail in the next chapter.

- **Country with maximum number of car listings**: First, select a list of count of cars grouped by country and then register it as a `temp` table. We call this temp view as `CAR_GRP_BYCNTRY`. Now, we fire a query on top of this view `CAR_GRP_BYCNTRY` to select the max count and from the max count figure out the country shown as follows:

```
DataSet<Row) grpByAggDF = spark.sql("select make_country, count(
*) as cnt
from cars group by make_country order by cnt desc");
grpByAggDF.createOrReplaceTempView("CAR_GRP_BYCNTRY");

DataSet<Row) countryWithMaxCarDF = spark.sql("select n.make_
country,n.cnt
From CAR_GRP_BYCNTRY n, (select max(cnt) as m from CAR_GRP_BY_
CNTRY)
c where n.cnt = c.m");

countryWithMaxCarDF.show();
```

This will print the result as follows:

- **Saving data to external storage**: The results obtained via Spark SQL queries can easily be dumped into external storage for future use. You can re-read the external stored files and build the dataframes again and fire queries on top of them.

- **Saving to file (as JSON or Parquet)**: As discussed earlier, Spark helps us store/read the content in various formats. Here I will show storing the results in JSON or Parquet format.

    1. **Select the cars from Italy and save them to external storage as JSON**: First we load the json dataset. Once the dataset is loaded, we register it as a temporary view. Now we fire our queries on that and create a new dataset. Finally, we can dump this dataset into an external storage using the format of data we want to use.

        ```
        Dataset<Row> carsBaseDF =
        spark.read().json("src/resources/data/cars.json") ;
        carsBaseDF.createOrReplaceTempView("cars");

        Dataset<Row> italyCarsDF =
        spark.sql("select * from cars where make_country='Italy' ");

        italyCarsDF.write().format("json").save("C:/temp/italycars") ;
        ```

As shown in the last line of the preceding code, we specify the format of storage that is, `json` and save it to an external directory. If you go to this external directory, you will see that there is a folder named `italy cars` and within that there will be a file starting with the word 'part' (that is, this depicts the partitioned data; in case of large datasets, the data is partitioned into multiple files). Some of the lines from this data are shown as follows:

```
{"make_country": "Italy","make_display":"Abarth","make_
id":"abarth","make_is_common":"0"}
{"make_country":"Italy","make_display":"AlfaRomeo","make_
id":"alfaromeo","make_is_common":"1"}
{"make_country":"Italy","make_display":"Autobianchi","make_
id":"autobianchi","make_is_common":  "0"}
```

Take a look at the `make_country` attribute in the preceding code, all are Italian cars.

2. **Save as parquet**: We can also store the data we stored earlier as JSON in other formats. This is just a simple change. We will change the format of the storage to `parquet` and the rest of the code remains the same:

```
italyCarsDF.format("parquet").save("resources/temp/pqt/
italyData");
```

When we store data to an external directory, it does not matter from a big data perspective whether the directory is on a filesystem or the directory belongs to a place in HDFS. In fact, in real-world applications, you will be partitioning and storing the data mostly on HDFS in some form (for example, as parquet or JSON, and so on).

> Hadoop runs on various other filesystems such as Amazon S3, so the output files can be saved to these filesystems as well.

3. **Saving to HDFS**: The files that we are saving on the operating system filesystem can also be pushed to HDFS or to any third-party filesystem such as Amazon S3:

   Here we will see how we can save the dataframe on HDFS:

```
italyCarsDF.write().format("parquet").save("<PATH_IN_
HDFS>");
```

If you execute the preceding code from a machine that is on big data stack and uses HDFS filesystem, it will then create and insert the data on HDFS.

4. **Re-read the stored data:** Let's now re-read the Italian cars data that we stored to external storage earlier. We will select and print the Italian cars data but this time read the data from external storage that is the external JSON file `italyData.json`. This file can also reside on HDFS too, apart from the normal filesystem:

```
DataFrame newItalyCarsDF = 
sqlCtx.read().format("json").json("resources/temp/italyData.json");
newItalyCarsDF.registerTempTable("italy_cars");
DataFrame italyCarsDF = sqlCtx.sql("select * from italy_cars");
italyCarsDF.show();
```

As you can see in the preceding code, we just loaded our data back from the external JSON file. We registered the dataframe as a temporary view and fired another query on it too.

Spark SQL for data exploration and analytics

Whatever we have depicted earlier using Spark SQL is a simple form of analytics that can both be used in real-world analytics as well as for exploring your data. So you can easily use Spark SQL and run queries for counting your data, finding distinct values or grouping your data to find counts, and so on by categories. Even though these are simplistic tasks, yet they are very powerful and in many use cases perhaps these are the only analytics pieces you might need.

Next let's look into a more complex analytics problem. In this problem we will try to analyze what is in the shopping basket of a consumer and based on that we will build some deductions and rules.

Market basket analysis – Apriori algorithm

When we shop at any store, we get a receipt of all the items we bought. This receipt is one transaction and it can have a very unique ID called transaction ID in the shopping store's database. Note that the store can be an online e-commerce store too. They keep all these transactions in a database to later study them.

This transaction history is valuable information for the shop owners or the e-commerce stores. It tells them about the buying patterns of the customers. Using this information, they can figure out which items sell the most, or which items go together. This will help them to arrange items accordingly in the different isles in their shop. For example they can keep chocolate cookies near to the isle containing milk as they know that lots of people who buy milk generally tend to buy chocolate cookies too. Similar to this, an online store can display items that go together, as shown next from one sample ecommerce store:

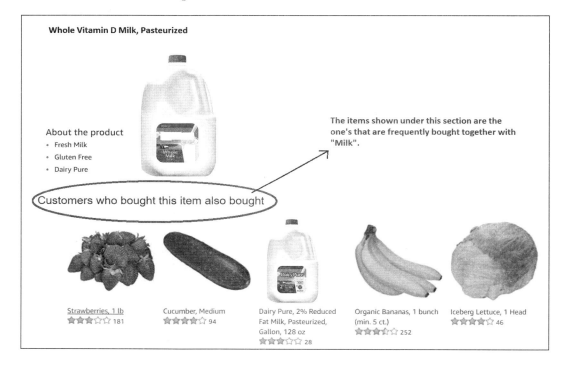

Suppose we get a list of few such transactions containing users buying different items, as shown in the following table:

Transaction ID	Items bought (in each transaction)
12761	Chocolate cookies, milk, and papaya
32343	Apples, milk, and diapers
43787	Chocolate cookies, apples, milk, and diapers
77887	Apples and diapers

As you can see there are four transactions in the preceding table.

Now, let's try to analyze this transaction set and while analyzing we will also study a very popular data mining algorithm called Apriori algorithm. We will go step-by-step through this evaluation:

- **Item frequency**: Find the frequency of each item within this transaction list. Frequency is nothing but the number of times this item is bought within this set of transactions. The values are shown as follows:

Item	Number of times it was sold	Support
Milk	3	¾ = 0.75
Chocolate cookie	2	2/4 = 0.5
Apples	3	¾ = 0.75
Papaya	1	¼ = 0.25
Diapers	3	¾ = 0.75

From this count of number of items sold, we can see that milk, diapers, and apples are each sold three times and papaya is sold only once. Thus, papaya is quite an infrequent item and is not sold much; hence, it does not look important for the purpose of analysis at all.

But what does the data in the last column under **Support** mean?

For studying the Apriori algorithms and its analysis, we must learn a few concepts. Let's try to understand these concepts now:

- **Support**: In simple terms, support just shows the ratio of the number of times a particular item or set of items is sold divided by the total number of transactions as shown:

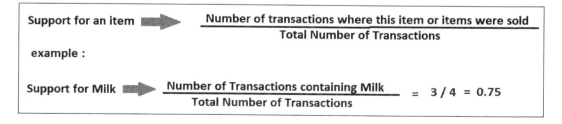

- **Minimum support**: If you look at the preceding set of transactions, you can easily see that there are a lot of combinations for items. So milk goes with chocolate cookies, milk goes with apples, chocolate cookies go with apples, and so on. As you can see, there are lots of combinations that you can build based on this small transaction history dataset itself. But an actual store has lots of items. If we try building this combination of items across the whole product list of the store (based on their transactions), you can easily make out that the amount of combinations and calculations would soon reach to an unmanageable quantity.

  To facilitate in reducing the number of combinations to analyze, the Apriori algorithm asks the users to set a minimum support ratio below which they can discard the item or item sets with that support level.

  Thus, if the minimum support is 0.5 (that is, the item should be present in minimum of half of the transactions), then we can discard the items that do not meet this minimum support value and create the most frequent item set from the previous set shown as follows:

Item	Number of times item was sold	Support	Description
Milk	3	¾ = 0.75	Selected as the value is greater than minimum support
Chocolate cookie	2	2/4 = 0.5	Selected as the value is greater than minimum support
Apples	3	¾ = 0.75	Selected as the value is greater than minimum support
Papaya	1	¼ = 0.25	Discarded as the value is less than minimum support
Diapers	3	¾ = 0.75	Selected as the value is greater than minimum support

As you can see, we discarded Papaya as it was mentioned in only one transaction and its support value is lower than the minimum support value. But this brings up an important question.

What is the rule behind choosing only frequently sold items or items that are frequently sold together?

Here comes an important rule from the Apriori algorithm, which states that, "If an item set is frequent then its subsets will be frequent too". As an example, if milk and apples are mentioned frequently in many transactions, then their subsets, that is, just apples and just milk are also mentioned in a lot of transactions and hence are frequent too.

- **Association rule**: Before we dig deeper into this analysis approach, let's try to understand what an association rule is.

    An association rule is an if...then...else type of statement that will link unrelated data within a database. So if we say "If we buy an item A, then we are most likely to buy another item B", then this is an association rule.

    Association rules are written as follows:

    $A => B$

    That is, if the **Left Hand Side** (**L.H.S**) is present, then Right Hand Side (**R.H.S**) is most likely going to be present.

    An example of an association rule could be as follows:

    {Chocolate cookies, Milk} => {Apples}

    Thus, if somebody buys chocolate cookies and milk, then they are likely to buy apples.

    If we have a transaction dataset of millions of transactions with thousands if not millions of items sold, we will have a huge number of association rules. Managing such a vast number of rules is a waste of computation effort, as not all rules will hold good, so let's now try to find how good our association rules are.

    How good is an association rule?

    From the preceding transaction dataset based on different combinations of items, we can figure out a lot of association rules as shown earlier. But do we know which ones are really good enough and which we can utilize?

    To figure out which rules are good enough for us to use, we use the concept of confidence. Thus, we will try to figure out how much confidence we have in a rule.

- **Confidence**: Confidence is the measure of goodness of our association rule. Thus, for a given association rule *{A} => {B}* within a set of transaction is defined as the proportion of transactions that contain *A* and also contain *Y*.

    Thus, for *{A} => {B}*, the confidence value is as shown :

> Support of {A} U {B} shows the value of support from transactions where both A and B are found.

Let's try to calculate confidence for the following rule:

{Chocolate cookie, Milk} => {Apple}

The confidence for this rule will be calculated using the following three steps:

1. First we calculate the support for chocolate cookies, milk, and apples as:

    Support of *{Chocolate cookie, milk, apples}* = ¼ = 0.25

2. Then we calculate the support for chocolate cookies and milk as:

    Support of *{Chocolate cookie, Milk}* = 2/4 = 0.5

Finally we calculate the confidence for our association rule as:

Confidence for *{Chocolate cookie, Milk} => {Apple}* = 0.25/0.5 = 0.5

Converting this real value result of *0.5* to a percentage we get the value as 50 percent. Thus we have 50% confidence on this association rule.

Full Apriori algorithm

In the previous steps, we saw how item frequency is found and we went over the concepts of how support, minimum support, and confidence is calculated. Now, let's look at the full Apriori algorithm given that we have the minimum support level that we want to run this algorithm on. Apriori algorithm comprises the following steps:

1. First find the frequent items and item sets.
2. Discard the item sets that have a frequency lesser than our minimum support level.
3. Figure out the association rules from these item sets and figure out their confidence levels.
4. Discard the rules that have confidence lesser than the value we are looking for and sort the association rules in descending order with values with higher confidence listed on top.

We will now put these steps into action and walk through an entire Apriori implementation on a sample dataset that we showed earlier:

- **Dataset**: Let's get back to our example dataset and solve the full problem now and build the association rules. As you must have noticed, we had put a minimum support of 0.5. After removing the items that did not meet the minimum support, we got a frequent item set shown as follows.

Item	Number of times it was sold	Support
Milk	3	¾ = 0.75
Chocolate cookie	2	2/4 = 0.5
Apples	3	¾ = 0.75
Diapers	3	¾ = 0.75

- **Apriori implementation**: After we collect the single items, we form the subsets of these items by combining them and forming combinations. The combinations can be shown as follows:

   Our individual items are => { chocolate cookie, milk , Apples, diapers }

   From these items, we can now make combinations as *{ chocolate cookie, milk }, { chocolate cookie, apples }, { apples, diapers }*, and so on.

The full list is shown in the following table. We also collect the Support for these item sets. Support will be the transaction containing these item sets divided by the total number of transactions:

Item sets	Number of times these item sets are seen in the transactions	Support
Chocolate cookies and milk	2	2/4 = 0.5
Apples and milk	2	2/4 = 0.5
Diapers and milk	2	2/4 = 0.5
Apples, chocolate cookies	1	¼ = 0.25
Chocolate cookies and diapers	1	¼ = 0.25
Apples and diapers	2	2/4 = 0.5
Apples, chocolate cookies, and milk	1	¼ = 0.25
Chocolate cookies, diapers, and milk	1	¼ = 0.25
Apples, chocolate cookies, diapers, and milk	1	¼ = 0.25
Apples, diapers, and milk	1	¼ = 0.25
Apples, chocolate cookies, and diapers	1	¼ = 0.25

As shown, we will reject all the item sets or combinations that do not meet our minimum support value. In our case, we use the minimum support value as 0.5 hence some of the item sets (in grey color in the preceding table) are rejected.

So now our set of combinations that passed our minimum support value will be as follows:

Item sets	Number of times these item sets are seen in the transactions	Support
Chocolate cookies and milk	2	2/4 = 0.5
Apples and milk	2	2/4 = 0.5
Diapers and milk	2	2/4 = 0.5
Apples and diapers	2	2/4 = 0.5

From these combinations, we now form the association rules. As we said earlier, the association rule is like an `if...else` statement, which states that if the left hand side happens then the right hand side might happen, that is:

'if somebody bought apples' => *'they might buy milk.*

We can write this rule as:

{apples} => {milk}

As we can see, this rule denotes if L.H.S happens, then R.H.S is possible.

But how do we know that the rule that we have depicted here is good enough?

Enter the confidence value that we had explained earlier. So how do we find our confidence in the preceding rule? We will use the following formula:

Support of L.H.S and R.H.S both   = Support for both Apples and Milk

Support of L.H.S only   Support for Apples only

Result =   (2/4)   =   2/3 = 0.66

(¾)

Thus, our confidence in this rule *{Apples} => {Milk}* is 0.66 or 66%.

Now, let's see all the rules based on the combinations we selected earlier:

Left Hand Side (LHS)	Right Hand Side (RHS)	Confidence
Chocolate cookies	Milk	2/4 = 1 or 100% 2/4
Milk	Chocolate cookie	2/4 = 0.66 3/4
Apples	Milk	0.66
Milk	Apples	2/4 / ¾ = 0.66
Diapers	Milk	(2/4) / (¾) = 0.66
Milk	Diapers	¾ / ¾ = 1 (this is 100%)
Apples	Diapers	¾ / ¾ = 1
Diapers	Apples	¾ / ¾ = 1

As you can see, the minimum confidence in our rules is 66% or 0.66.

What if the user says that they are only interested in rules with a minimum confidence level of 80% ?

In this case, we will filter out the rules that have lesser confidence than this, and we will have the following rules from the preceding tables.

Note, here we will write the rules in the proper format.

So our final result is as follows:

Rule	Confidence
{ Chocolate cookies, Milk }	1
{ Milk, Diapers }	1
{ Apples, Diapers }	1
{ Diapers, Apples }	1

Implementation of the Apriori algorithm in Apache Spark

We have gone through the preceding algorithm. Now we will try to write the entire algorithm in Spark. Spark does not have a default implementation of Apriori algorithm, so we will have to write our own implementation as shown next (refer to the comments in the code as well).

First, we will have the regular boilerplate code to initiate the Spark configuration and context:

```
SparkConf conf = new SparkConf().setAppName(appName).
setMaster(master);
JavaSparkContext sc = new JavaSparkContext(conf);
```

Now, we will load the dataset file using the `SparkContext` and store the result in a `JavaRDD` instance. We will create the instance of the `AprioriUtil` class. This class contains the methods for calculating the support and confidence values. Finally, we will store the total number of transactions (stored in the `transactionCount` variable) so that this variable can be broadcasted and reused on different DataNodes when needed:

```
JavaRDD<String> rddX = sc.textFile(FILE_NAME);
AprioriUtil au = new AprioriUtil();
Long transactionCount = rddX.count();
Broadcast<Integer> broadcastVar = sc.broadcast(transactionCount.
intValue());
```

We will now find the frequency of items. By frequency we mean the number of times the item and its combination with other items is repeated in the transactions. The `UniqueCombinations` class instance contains the utility methods for helping us find the item combinations, which we later use to find their frequency in the dataset of transactions. As shown in the following methods, we first find the combinations and later use the combination values as the key so that we can run `reduceByKey` operations to sum up their frequency count as follows:

```
UniqueCombinations uc = new UniqueCombinations();
JavaRDD<Map<String,String>> combStrArr = rddX.map(s ->
uc.findCombinations(s));
JavaRDD<Set<String>> combStrKeySet = combStrArr.map(m -> m.keySet());
JavaRDD<String> combStrFlatMap = combStrKeySet.flatMap((Set<String> f)
->
f.iterator());
JavaPairRDD<String, Integer> combCountIndv = combStrFlatMap.
mapToPair(s -> new Tuple2(s, 1));
JavaPairRDD<String, Integer> combCountTotal = combCountIndv.
reduceByKey((Integer x, Integer y) -> x.intValue() + y.intValue());
```

Now we will collect the items and their count as well as the item combinations and their count and store them in a `Map` for future use within the program. To make the collection available across different DataNodes, we put this `Map` in a `Broadcast` variable:

```
Map<String,Integer> freqMap = combCountTotal.collectAsMap();
Broadcast<Map<String,Integer>> bcFreqMap = sc.broadcast(freqMap);
```

Now we will be filtering items with frequency less than support:

 Support is the number of minimum counts or frequency of an item.

```
JavaPairRDD<String,Integer> combFilterBySupport = combCountTotal.
filter(c ->
c._2.intValue() >= 2);
```

Since we are interested in association rules where we want to depict that if one item or items are present then another item might also be present, we are interested in a combination of items only; hence, we will filter our rules with just a single item in it:

```
JavaPairRDD<String,Integer> freqBoughtTogether = combFilterBySupport.
filter(s ->
s._1.indexOf(",") > 0);
```

Let's start building the actual association rules now. For this, we will go over the items of frequently bought together RDD and we would invoke a `flatMap` method on it. This `flatMap` function would break the individual rows into a collection of objects. On this collection, we will invoke a method `getRules` from our `UniqueCombinations` class. This method `getRules` would break the items into left hand side and right hand side combinations and store in a `Rule` object. Finally, we will figure out the support of the left-hand side value, support of the right-hand side value and the confidence value of this rule and store the rule in a collection and return the result in RDD:

```
JavaRDD<Rule> assocRules = freqBoughtTogether.flatMap(tp -> {
List<Rule> rules = uc.getRules(tp._1);
  for (Rule rule : rules) {
  String lhs = rule.getLhs();
  String rhs = rule.getRhs();
  Integer lhsCnt = bcFreqMap.value().get(lhs);
  Integer rhsCnt = bcFreqMap.value().get(rhs);
  Integer lhsRhsBothCnt = bcFreqMap.value().get(tp._1);
  double supportLhs = au.findSupport(lhsCnt, broadcastVar.value());
  double supportRhs = au.findSupport(rhsCnt, broadcastVar.value());
  double confidence = au.findConfidence(lhsRhsBothCnt, lhsCnt);
        rule.setSupportLhs(supportLhs);
          rule.setSupportRhs(supportRhs);
          rule.setConfidence(confidence);
  }
  return rules.iterator();
});
```

We will start printing our association rules now:

 At this point, we can also filter out the rules that do not meet our minimum **confidence** criteria.

```
List<Rule> rulesColl = assocRules.collect();
for (Rule rl : rulesColl) {
    System.out.println(rl.getLhs() + " => " + rl.getRhs() + " , " +
rl.getConfidence());
}
```

We will now see the uses and any disadvantages of this algorithm:

- **Use of Apriori algorithm**: Apriori can be used in places where the number of data transactions is small. It's a simpler algorithm and is easy to maintain on small amounts of data.
- **Disadvantages of Apriori algorithm**: Even though Apriori algorithm is easy to code and use, it has some disadvantages:
    - The main disadvantage of Apriori algorithm is that it is slow. For such a small dataset as we used earlier there were so many combinations. Thus, on a very large dataset, it can generate millions of combinations and so computation-wise it can be slow.
    - If the number of counts of items sold or combinations sold increases (which will happen as items do get sold), then the algorithm will have to rescan the entire dataset and do the computation again.

 Thus, as you can see, Apriori algorithm is a good choice for smaller datasets.

Efficient market basket analysis using FP-Growth algorithm

The Apriori algorithm is slow and requires lot of computation power. When the number of transactions is very high, the item combination count explodes and becomes too expensive to compute. Hence, Apriori is not a practical approach on very large datasets.

Chapter 2

To avoid the pitfalls in Apriori, the FP-Growth algorithm was developed. This algorithm is especially suited for big data operations and goes well with Apache Spark and MapReduce. Spark comes with a default implementation of FP-Growth algorithm in its MLlib library.

Let's now try to understand the concepts behind the FP-Growth algorithm.

What is a FP-Growth algorithm?

FP-Growth algorithm builds on top of the Apriori algorithm and is essentially an improvement on top of it. It avoids the pitfalls of the Apriori algorithm and is very fast to run on large datasets. The FP-Growth algorithm uses a different approach than Apriori and reads the databases of transactions only twice as compared to Apriori (which has to read the database multiple times) and hence it is much faster. The algorithm reads through the dataset of transactions and creates a special data structure called FP-Tree. An example of a Dtree is shown next:

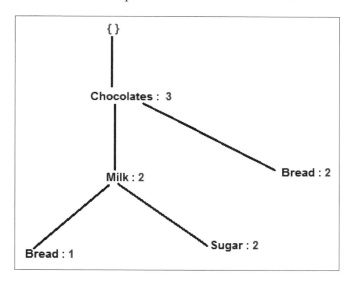

As you can see in the preceding diagram, **Chocolates** has the number **3** next to it. This number depicts the number of transaction encountered until that particular row of the dataset.

 Unlike normal search trees, the Node in the Dtree can be repeated As you can see in the preceding diagram, the node **Bread** is repeated in this tree.

[55]

First Steps in Data Analysis

The approach to using the FP-Growth algorithm for market basket analysis can be described in two steps:

- Build the FP-Tree
- Find frequent item sets using this FP-Tree

Let's now go through the full FP-Growth algorithm step-by-step:

- **Transaction dataset**: So let's suppose we get a list of a few transactions containing users buying different items shown in the following table:

Transaction ID	Items bought in each transaction
12761	Chocolate cookies, milk, and papaya
32343	Apples, milk, and diapers
43787	Chocolate cookies, apples, milk, and diapers
77887	Apples and diapers
77717	Milk and oranges

- **Calculating the frequency of items**: Now, let's find the frequency of each item in the datasets. Frequency is nothing but the number of times the items show up with in the set of transactions. For example, apples show up three times in the preceding transactions, hence the frequency is three. The full set of values is shown in the following table:

Item	Frequency
Chocolate cookie	2
Milk	4
Papaya	1
Apples	3
Diapers	3
Oranges	1

- **Assign priority to items**: Now assign a priority to each item in the preceding list. The item which has more frequency in the dataset is given a higher priority. The priority of the items are shown in red color in the following table:

Item	Frequency	Priority
Chocolate cookie	2	4
milk	4	1
papaya	1	5
Apples	3	2
diapers	3	3
oranges	1	6

- **Array items by priority**: Now we go back to the transaction list we had in 10.1 and arrange the items in the order of priority with the highest priority items coming first and lowest one in the end as shown in the following table:

Transaction ID	Items bought	Items according to priority
12761	Chocolate cookie, milk, and papaya	Milk, chocolate cookie, and papaya
32343	Apples, milk, and diapers	Milk, apples, and diapers
43787	Chocolate cookie, apples, milk, and diapers	Milk, apples, diapers, and chocolate cookie
77887	Apples and diapers	Apples and diapers
77717	Milk and oranges	Milk and oranges

- **Building the FP-Tree**: Once you have arranged the items in the sort order according to priority, it is the time to build the FP-Tree.

  We will start building the FP-Tree by going from the sorted list of first transaction items downward until we cover all the transactions. The steps are listed as follows:

  1. The first transaction is {`Milk, chocolate cookie, papaya` } (in the order of priority).

First Steps in Data Analysis

For FP-Tree, the initial node is always null or blank and the remaining nodes root out from it. While writing the item in the node, we write the count of its occurrence until that transaction. So since {Milk, chocolate cookie, papaya } is the first transaction only, the occurrence count of all the items will be 1 until this transaction and the tree would be as shown next:

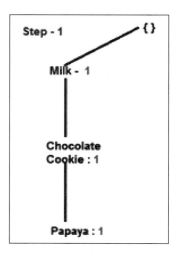

As you can see in the preceding diagram, the FP-Tree starts with a null node and at each node element you can see the count of that element (up to that transaction).

2. Now pull the second transaction and take the items and see if you need to walk through the same path of the tree; if not, then you create a new path. If you walk through the same path, then you increase the item counts in the entry that already exist. If you are going on a new path in the tree, then you put the item count as if it's the first transaction:

Chapter 2

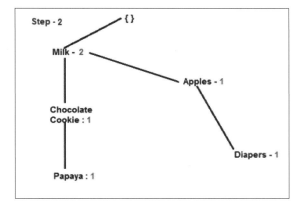

As you can see in the preceding diagram, **Milk** was on the same path twice so we increased its value by 1 whenever it occurred across the same path in a different transaction. Also, we treat items on a new path as new items and add a new count of 1 to them. On the new path, the items can be repeated.

3. Now we pick the third transaction and plot it on the tree. Note that none of the existing paths start with **Chocolate Cookie**; hence, we create a new path. This new path will contain items that we already mentioned in the FP-Tree earlier. This is the main difference between an FP-Tree and a normal search tree. It can contain repeated items as we mentioned earlier:

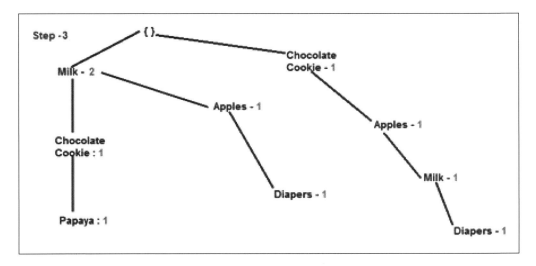

4. **Remaining steps**: Plotting the remaining transactions on the tree will show the full FP-Tree as shown next:

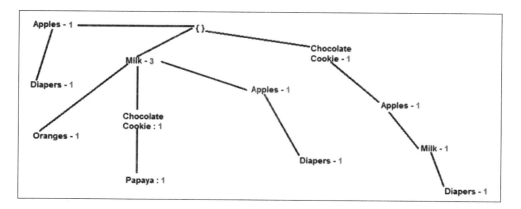

 {Milk, Oranges} starts with the existing node **Milk**, so we increase its count further to **3**.

As you can see, the whole set of transactions is now fitted into one small tree. In a way, FP-Tree can compress a huge list of transactions into a sorted tree structure. This tree structure in some cases can be easily fitted into a computer's memory for fast computations. In the case of big data, huge FP-Trees can be distributed across a cluster of machines.

We have seen the full FP-Tree now but we need to find out a few more details about the tree as mentioned next.

How do you validate whether the FP-Tree that you have built is good or not?

The count of the total number of items should exactly match the count of the items in the original transaction. Thus, as you can see in the preceding diagram, the count of **Milk** is **4**.

 Count of Milk Items from the FP-Tree is = Count on First Milk Node + Count on Second Milk Node = 3 + 1 = 4.

Hence, this is correct as it matches the original count of **Milk** transactions from the dataset.

What about the count of **Diapers**?

As you can see, the diaper's node is mentioned 3 times in the preceding diagram. Hence, the count of diapers is 3 and this matches the original count of diapers in the transaction dataset.

- **Identifying frequent patterns from the FP-Tree**: Before we find the frequent patterns from the FP-Tree, let's check what the minimum support is on which we want to find the frequent patterns. Let's try to evaluate the pattern at a minimum support of 0.4.

  If the minimum support is *0.4*, then what will the support be? Let's see in the following formula:

  *Support = minimum support * Number of total transactions*

  Therefore, *Support = 0.4 * 5 = 2*

  After we have the *minimum support* or *support value* we can start mining the conditional patterns from the FP-Tree we built earlier.

- **Mining the conditional patterns**: To mine the conditional patterns, we will go over the tree recursively from the leaf nodes onwards, that is, from the bottom of the tree upwards. However, the path that we will mention and use for patterns will always be from the top of the root node to the node we are evaluating.

  Let's see the Full FP-Tree again along with the frequencies of each item:

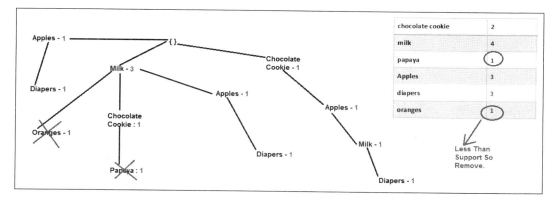

We will traverse the tree from the bottom upwards, but first we reject the items whose frequencies are lesser than our minimum support value, that is, **2**.

Hence, as shown in the preceding diagram (in red crosses), **Oranges** and **Papaya** are rejected.

Now, let's start building the conditional pattern base and conditional FP-Tree.

Let's start with each item and traverse the tree upwards from the leaf nodes and find the conditional patterns. We will pick one node Diaper and traverse up and we leave it up to our readers to traverse remaining nodes and figure out the same. The approach would be exactly the same as we'll explain for the node Diaper. The reader can later match their results with the result that our actual Apache Spark FP-Growth program produces.

> As I also mentioned earlier, for the purpose of explaining this, I am only using the leaf node **Diapers**. For brevity in this section of the book, I leave it to the readers to do the same for other items in the dataset.

- **Conditional patterns from leaf node Diapers**: The leaf node **Diapers** can be reached by the following paths as shown in the following figure:

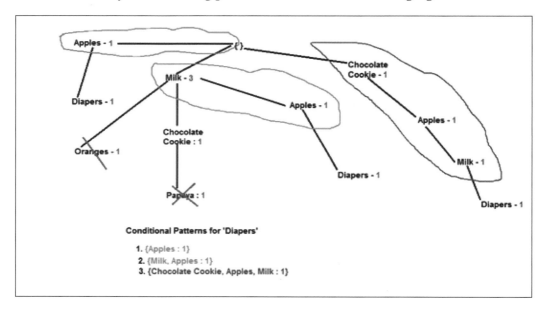

Conditional Patterns for 'Diapers'

1. {Apples : 1}
2. {Milk, Apples : 1}
3. {Chocolate Cookie, Apples, Milk : 1}

In the preceding figure, the blue, green, and purple lines denote the various paths by which **Diapers** can be reached in the FP-Tree and the number of its occurrences within that path. Each path denotes a conditional pattern. Hence, all the conditional patterns for diapers are:

Item	Conditional patterns
Diapers	{ Apples : 1}, {Milk, Apples : 1}, {Chocolate Cookie, Apples, Milk : 1}

Now that you have the conditional patterns, let's look at our next question:

How will you make the conditional tree out of the conditional pattern you found for diapers?

Making the conditional tree is simple. Just use the conditional patterns as a new dataset (given the condition that diapers are already selected). So, with this new dataset, we'll now make the conditional tree, but first we'll make the header table with the item frequencies and priorities.

Item	Frequency (or count)	Priority
Apples	3	1
Milk	2	2
Chocolate cookie	1	3

Now, let's make the tree again iterating the transactions:

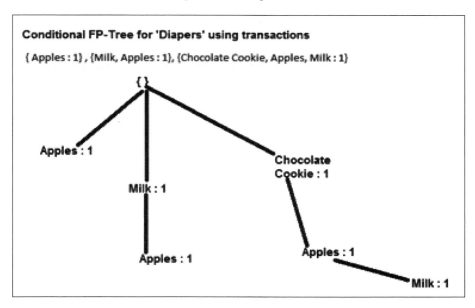

As you can see, this is our Conditional FP-Tree for **Diapers** based on the conditional patterns we found for it.

First Steps in Data Analysis

Now, let's again build the **Conditional Patterns Base**, **Conditional FP Tree**, and **Frequent Item** sets from the tree shown in the preceding table:

Item	Conditional Pattern	Conditional FP-tree	Frequent item sets	Meets 'support'
Apples	{Milk : 1}, {Chocolate Cookie : 1}		-	None of the items meet minimum support
Milk	{Chocolate Cookie, Apples : 1 }	-	-	None of the items meet minimum support
Chocolate Cookie	No Items	-	-	No items

As all the **Conditional Pattern** items below the minimum support count are rejected, there are no frequent item sets from the **Conditional-FP tree**. Here, we just checked the combination and rejected the combinations, but we have not evaluated the individual items yet. The individual items from the conditional pattern of diapers had some values greater than the minimum support as shown in the following table:

Item	Frequency (or Count)	Priority	Meets 'Support'
Apples	3	1	Yes
Milk	2	2	Yes
Chocolate cookie	1	3	No

 We have eliminated all the items that did not meet the minimum support here as well.

As you can see, only two items are above the minimum support and they are {Apples : 2} and {Milk : 2}.

Thus, we get the frequent item set for diapers as shown next:

{Apples, Diapers : 3}, {Milk, Diapers : 2} and the combination of apples and milk along with diapers, that is, {Apples, Milk, Diapers}.

So, finally, we have the list of association rules for the item **Diapers**. But our algorithm is not finished yet. We need to repeat the same approach for the other items. Similar to what we did for diapers earlier, we have to go over all the other items in the list of our main Header Table, that is, our original list of items. For refreshing your memory, I have shown the same list of transactions again:

Items from main transaction dataset	Frequency	Priority
Milk	4	1
Apples	3	2
Diapers	3	3
Chocolate cookie	2	4

As shown in the preceding table, we have the original list of items and their frequency. We have also reshown the priority here. From this we find **Conditional Patterns**, **Conditional FP-Tree**, frequent itemsets for the remaining items, that is, chocolate cookie, apples, and milk in a similar way as we did for **Diapers**.

I have discarded oranges and papaya in the preceding table as they don't meet the minimum support. This is our assumption that if the items are infrequent, their combinations would be infrequent too, hence they are completely discarded.

FP-Growth algorithm has some clear advantages over Apriori; we will list the following advantages below:

What are the advantages of using FP-Growth over Apriori?

FP-Growth is a very popular algorithm and it is much more popular than Apriori because:

- It only requires you to scan the transaction dataset two times for building the FP-Tree. On the other hand, in Apriori, you have to scan the transaction dataset again and again. If the number of transactions in the dataset is very high, which is quite possible for a big data project, then Apriori will become just too slow to handle. On the other hand, FP-Growth will be much faster in this case.

- FP-Growth compresses the transaction data in the form of an FP-Tree data structure; hence, it can fit in memory too in certain cases if the data is good enough to fit in memory. In this case, it will become even faster.

First Steps in Data Analysis

- There is a parallel version of FP-Growth algorithm (this is the version that Spark uses). It distributes the FP-Growth computation on a cluster of machines and this version is massively scalable.

Running FP-Growth on Apache Spark

Apache Spark implements a parallel version of FP-Growth called as **PFP**. In this version, the dataset of transactions is broken and distributed across a cluster of machines. So the frequency count of operations is individually done on the cluster of machines. Later the result of the frequency count is combined. This algorithm now groups the transactions into different groups. The groups are individually independent in such a way that FP-Trees can be locally built based on them on different machines of the cluster. Later, the results can be combined for the frequent itemsets. Thus, this implementation is massively scalable. For a complete description of the algorithm, refer to the research paper mentioned on Spark documentation at https://spark.apache.org/docs/latest/mllib-frequent-pattern-mining.html. Even though the algorithm is distributed underlying the principle of finding frequent item sets the technique is still the same.

The Spark Java code for the same algorithm is shown next.

We are using the same transaction dataset as we used in our FP-Growth example earlier.

We will build the `SparkContext` with instance `sc`. Next, using this instance we load our dataset of transactions:

```
JavaRDD<String> data =
sc.textFile("resources/data/retail/retail_small_fpgrowth.txt");
```

Now, break each row of transaction into individual items and store these items as a list of strings per row in a Spark RDD object called `transactions`:

```
JavaRDD<List<String>> transactions = data.map(
    new Function<String, List<String>>() {
      public List<String> call(String line) {
        String[] parts = line.split(" ");
        return Arrays.asList(parts);
      }
    }
);
```

Now create an instance of FP-Growth algorithm that is provided by Apache Spark and is present in the MLlib API of Spark:

 On the FP-Growth instance, see how we are setting the value of the MinSupport as 0.4.

```
FPGrowth fpg = new FPGrowth().setMinSupport(0.4).setNumPartitions(1);
```

Now, run the FP-Growth algorithm instance on the transactions RDD object we built earlier. This would create the FP-Tree and create the association rules internally and store the results in a FPGrowthModel object:

```
FPGrowthModel<String> model = fpg.run(transactions);
```

Get the list of frequent items and print them. You can get this list from the FPGrowthModel instance you built earlier:

```
for (FPGrowth.FreqItemset<String> itemset:
model.freqItemsets().toJavaRDD().collect()) {
    System.out.println("[" + itemset.javaItems() + "], " + itemset.freq());
}
```

Now define the minimum confidence value and apply it to get the association rules from the FPGrowthModel object:

```
double minConfidence = 0.0;
for (AssociationRules.Rule<String> rule
      : model.generateAssociationRules(minConfidence).toJavaRDD().collect()) {
System.out.println(
   rule.javaAntecedent() + " => " + rule.javaConsequent() + ", " +
rule.confidence());
    }
}
```

And the results would be printed as follows:

```
[apples, milk] => [diapers], 1.0
[milk] => [diapers], 0.5
[milk] => [apples], 0.5
[milk] => [chocolate], 0.5
[apples] => [diapers], 1.0
[apples] => [milk], 0.66
[diapers, apples] => [milk], 0.66
[diapers] => [apples], 1.0
[diapers] => [milk], 0.66
[diapers, milk] => [apples], 1.0
[chocolate] => [milk], 1.0
```

Summary

We started this chapter on a simple note by going over the very basic yet very power simple analytics on simple datasets. While doing so, we also learned a very powerful module of Apache Spark called Spark SQL. Using this module, Java developers can use their regular SQL skills and analyze their big data datasets.

After exploring the simple analytics piece using `spark-sql`, we went over two complex analytic algorithms: Apriori and FP-Growth. We learned how we can use these algorithms to build association rules from a transaction dataset.

In the next chapter, we will learn the basics of machine learning and get an introduction to the machine learning approach for dealing with a predictive analytics problem.

3
Data Visualization

It's easier to analyze your data once you can view it. Viewing data requires putting your data points in charts or graphs that you can visualize and figure out the various details. You can also generate charts/graphs after running your analytic logic. This way you can visualize your analytical results as well. As a Java developer you have lots of open source tools at your disposal that you can use for visualizing your data and the results.

In this chapter we will cover:

- Six types of charts and their general use and concepts
- Sample datasets used in building the charts
- Brief JFreeChart introduction
- An example of each type of chart using the JFreeChart and Apache Spark API on big data

Data visualization with Java JFreeChart

JFreeChart is a popular open source chart library built in Java. It's used in various other open source projects as well such as JasperReports (open source reporting framework). You can build a number of popular charts such as pie charts, time series charts, and bar charts to visualize your data with this library.

JFreeChart builds the axis and legends in the charts and provides automatic features such as zooming into the charts with your mouse. For simple chart visualizations that the developer can use to build the models (using lesser data) JFreeChart is good but for extensive data visualization that you need to ship to your customers or end users you are better off with an elaborate data visualization product such as Tableau or QlikView over big data. Although we will cover some of the charts from JFreeChart, this chapter is by no means an extensive take on JFreeChart.

Data Visualization

For this book and its examples, we use these charts extensively for visualizing our datasets. In most of the cases, the boilerplate code (the code involved in building the chart) is the same and we will just change the dataset. From this dataset, we will pull the data and provide it in the format (read the type of dataset object) that the JFreeChart library requires and pass it to the chart object that we need. For most of our examples and practicing, this should be enough. For advanced data visualizations, we are better off using advanced data visualization tools such as Tableau or QlikView.

Using charts in big data analytics

It is said that a picture is worth a thousand words. In terms of big data, the amount of data is so high that by just plainly looking at raw data it is extremely difficult to figure out any trends in data. However, the same data when plotted on a chart is much more comprehensible and easier to identify trends or relationships within data. As an example, let's take a look at a simple time-series chart showing house prices versus year.

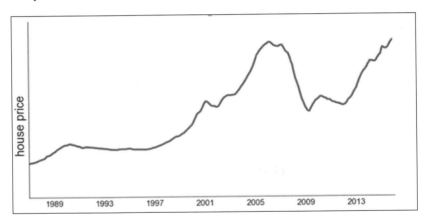

As you can see from the preceding chart, the house price almost kept rising after 1989 and reached its peak at around 2007, and after that it started falling before starting to rise again around 2009 onwards. It's not a bad analysis since 2007 was around the time we had the recession.

From the perspective of data analysis, charts are used mainly for two purposes:

1. **For initial data exploration**: Any raw data has to be explored first before it gets analyzed. Initial data exploration helps us in figuring out:
    - Any missing data
    - Null data
    - Erroneous data

- Outliers (erroneous or special points in data)
- It also helps us in making simple deductions from data such as count of rows in the dataset, average or mean calculations, percentiles, and so on

2. **For data visualization and reporting**: Making charts for exploration is mainly for the purpose of initial study of the data before it can be properly transformed for proper analysis via different algorithms. But you might have requirements to store the results of the data in different forms of reports or dashboards. This chapter mainly serves the purpose of data exploration and we do not dig into data reporting as it is a much bigger issue and beyond the scope of this book. For advanced data visualization and reporting, you can refer to tools such as FusionCharts, D3.js, Tableau, and so on.

Time Series chart

This is a simple chart used for measuring events over time or in other words it is a series of statistical observations that are recorded over time. Visualizing your data this way would help you figure out how the data changes with respect to time in the past and you can also make predictions regarding the values that might occur in the future when time changes. Let's now see some sample Time Series charts in action.

Before giving examples of time series charts, let's understand the dataset used for the time series chart examples.

All India seasonal and annual average temperature series dataset

In this dataset, we have India's seasonal temperature captured on monthly/annual basis from 1901 to 2015. The dataset is downloaded as a JSON file from `https://data.gov.in/catalog/all-india-seasonal-and-annual-mean-temperature-series`. You can also find the sample dataset in the GitHub code accompanied with this book.

Data Visualization

This dataset comprises two `json` objects as shown next:

- **Fields**: This `json` object contains the fields and labels for the data within the dataset. These are the fields present in the dataset as shown in the following table. I am only showing the indexes that we are using in the charts.

Table index	Label
0	Year
1	Temperature in Jan
2	Temperature in Feb
3	Temperature in March
4	Temperature in April
...	...
13	Average Annual Temperature

- **Data:** This `json` object contains the actual data in the form of a JSON array object. For simplicity, I have removed the `fields` object and put the `data` object in a single row so that it will be easy to process with Apache Spark. So, our data is of a single row as follows:

```
{"data":"1901","22.40","24.14","29.07","31.91","33.41","33.18","31.21"
...}
```

Simple single Time Series chart

Let's see a simple Time Series chart of average monthly temperature versus months in 2015. So, on the *y axis*, we will have the average temperature and on the *x axis* we will have the months of 2015 as shown in the following diagram:

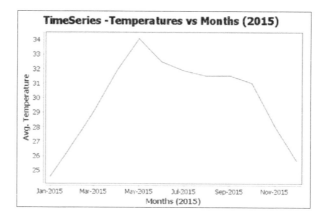

As you can see in the preceding diagram, the temperature spikes in the month of May which is quite true as these are very hot months in India. The graph also shows in general, the trend of the flow of data as you can also see the temperature keeps on going up until the mid of May and then it starts falling as the months get colder.

Now, let's create this chart via the JFreeChart library. Creating a JFreeChart chart requires some simple steps as follows:

1. Create a dataset in the format that is needed by the JFreeChart API:
    - **Load the dataset using Apache Spark**: Before starting any Spark program we build the Spark Context and Spark configuration first. So we will build the Spark config and `JavaSparkContext` object
        ```
        SparkConf sparkConf = new
                    SparkConf().setAppName("TimeSeriesExample")
         .setMaster("local");
                JavaSparkContext ctx = new
        JavaSparkContext(sparkConf);
        ```
        As we are going to use the Spark SQL queries to pull data from our dataset, for ease of usage we1 will next create the `SQLContext` object of Spark using the previously created `JavaSparkContext` object.
        ```
        SQLContext sqlContext = new SQLContext(ctx);
        ```
        After building our `SQLContext` object we are now ready to load our dataset from the dataset file. To the `sqlContext` object, we provide the format of the data that we are trying to load, in our case it is json format and we also provide the location of the file. The `sqlContext` object will load the data and store it in a dataset variable.
        ```
        Dataset<Row> df = sqlContext.read().format("json").
        json("data/india_temp.json");
                df.createOrReplaceTempView("india_temp");
        ```
    - Next, query the dataframe you created and pull all the data:
        ```
        Dataset<Row> dfc = sqlContext.sql("select explode(data) from india_temp");
        ```
    - Now, filter only the data for the year 2015:
        ```
        JavaRDD<Row> rdd = dfc.javaRDD();
        " " " "
        JavaRDD<Row> filterRdd = rdd.filter(s -> {
                if("2015".equals(s.getList(0).get(0).toString()))
        ```

Data Visualization

```
        return true;
        return false;
            });
```

- Finally, fill this data into a `TimeSeries` object and return the results:

```
final TimeSeries series=new TimeSeries("Jan-Dec2015");
List<Row> filterList = filterRdd.collect();
for(Row row : filterList) {
    List<String> items = row.getList(0);
    for(int i =1; i <=12; i++){
        series.add(new Month(i,
Integer.parseInt(items.get(0))),
new Double(items.get(i)));
    }
}
return new TimeSeriesCollection(series);
```

2. **Create the chart object**: As you can see in the following code, we build a Time Series chart by invoking `createTimeSeriesChart` on the `chartFactory` object and it is in this object that we also pass the chart name, *x axis* label, *y axis* label along with the dataset object itself. The library that we are using that is JFreeChart, extracts data from the dataset objects and starts building the chart using the other parameters specified in the method too.

```
chartFactory.createTimeSeriesChart("TimeSeries Temperatures
vs Months (2015) ","Months (2015)","Avg. Temperature",
dataset,false,false,false);
```

Finally, we have some boilerplate code that connects all this together. So, in this code, we build the dataset and pass it to the chart object. Finally, we add the chart object to the chart panel that is again pushed on the content pane.

```
final JFreeChart chart = createChart(dataset);
    chart.getPlot().setBackgroundPaint(Color.WHITE);
final ChartPanel chartPanel =new ChartPanel(chart);
chartPanel.setPreferredSize(new
java.awt.Dimension(560,370));
    chartPanel.setMouseZoomable(true,false);
    setContentPane(chartPanel);
```

For the full code of this example, refer to the code in the GitHub repository.

Multiple Time Series on a single chart window

If you show multiple charts in the same window, then you can easily visualize the comparison of variations of data over time. Here, I will show two Time Series charts in the same window. In one chart, we will see **Avg Temp** versus Months for 2014 and in the other chart we will see **Avg Temp** versus Months for 2015 as shown in the following diagram:

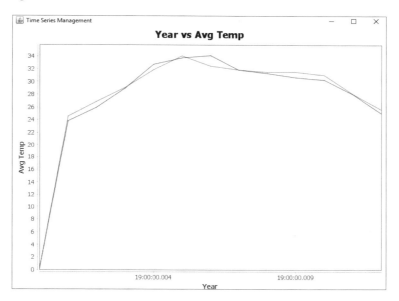

As you can see in the preceding diagram, it is easy to visualize the graphs for the years 2014 and 2015 and see how the temperature varies between the two. The blue line is the chart for the year 2014 and the red line is the chart for the year 2015.

For building any JFreeChart as shown in the examples in this book, most of the code needed to actually create the graph is almost boilerplate with a few configuration changes as needed. The main piece of the code is the `createDataset()` function. It is here that you need to pull the data from your data files or other sources of data and fill them into an object that the JFreeChart, which specific chart component needs. As long as the data is compliant with the dataset component that the chart requires, you can easily prepare the chart.

So, for the multiseries chart component, we would just change the `createDataset()` method. Let's now go through the code of this method.

Data Visualization

First, we create an instance of `DefaultXYDataset` and load the data using Spark. Once the data is loaded, we register the dataframe as a temporary view and query it to pull the data.

```
private XYDataset createDataset()
{
DefaultXYDataset ds = new DefaultXYDataset();
SparkConf sconf = new
SparkConf().setAppName(APP_NAME).setMaster(APP_MASTER
);
SparkSession spark = SparkSession.builder()
.config(sconf)
.getOrCreate();
Dataset<Row> df = spark.read().format("json").json(
"data/india_temp.json");
df.createOrReplaceTempView("india_temp");
Dataset<Row> dfc = spark.sql("select explode(data) from
india_temp");
        JavaRDD<Row> rdd = dfc.javaRDD();
```

Next, we filter the data for the years 2014 and 2015 using the following code:

```
JavaRDD<Row> filterRdd = rdd.filter(s -> {
        if("2015".equals(s.getList(0).get(0).toString()))
return true;
else
if("2014".equals(s.getList(0).get(0).toString()))
return true;
else return false;
        });
```

After the data for the years 2014 and 2015 is extracted, we fill this data in our `DefaultXYDataset` object and return it with the help of the following code:

```
List<Row> filterList = filterRdd.collect();
int j = 0;
for (Row row : filterList) {
double[][] series = new double[2][13];
List<String> items = row.getList(0);
for(int i = 1 ; i <= 12 ; i++) {
series[0][i] = (double)i;
    series[1][i] = new Double(items.get(i)) ;
}
ds.addSeries("Series-" + j, series);
j = j + 1;
}
return ds;
}
```

[76]

Bar charts

A bar chart shows variations in quantity of some entity using rectangles either drawn vertically or horizontally on a chart. As you visualize the different lengths of rectangles on the chart, it is easy to figure out which category is more and which one is less. Bar charts have three main advantages:

- You can see the data relationships in the x and y axes
- You can easily compare the values among different categories
- You can also use them to visualize trends

As an example, take a look at the following bar chart, which shows the number of cars made by different countries (as shown in `cars.json` dataset):

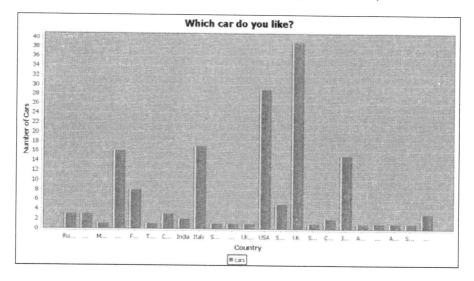

As you can see in the preceding chart, this dataset has a maximum number of cars from the UK followed by the USA, followed by Italy, and so on.

Let's explore this example further with the actual code. The `cars.json` dataset that is analyzed by the preceding chart, has the following format:

```
{"make_id":"abarth","make_display":"Abarth","make_is_common":"0","make_country":"Italy"}

{"make_id":"ac","make_display":"AC","make_is_common":"0","make_country":"UK"}
```

```
{"make_id":"acura","make_display":"Acura","make_is_common":"1","make_
country":"USA"}

{"make_id":"alfa-romeo","make_display":"Alfa
Romeo","make_is_common":"1","make_country":"Italy"}
```

This dataset contains information about cars from different countries. It has a `json` object per row and within the `json` object it has details for one particular car. Some of the main attributes within one `json` object are shown in the following table:

Attribute name	Description
`make_display`	Display name of car, the name used in 'Ads'.
`make_country`	Country where the car is made
`make_id`	Name of car or model name

To build the bar chart, we follow similar steps to what we discussed in the preceding Time Series chart. Let's see how to do it:

1. **Create the dataset**: We will load the `cars.json` file using Apache Spark and build a dataset object (from JFreeChart library). Here is the code for that (before we write the main code for our Spark program we build the `SparkSession` object)

    ```
    SparkSession spark = SparkSession
      .builder()
      .appName("Learning charts for
         analytics")
      .config("spark.some.config.option", "some-value")
      .getOrCreate();
    ```

    Now we have our `SparkSession` object ready. Using this `SparkSession` object, we next go on to load our dataset from the dataset file.

    First we load the dataset `json` file into a dataframe and register it as a temporary view:

    ```
    Dataset<Row> cdf =
    spark.read().format("json").json("data/cars.json");
            cdf.createOrReplaceTempView("cars");
    ```

Next we query it to pull the temp view to pull the countries and the number of cars in each country:

```
Dataset<Row> cdfByCountry = spark.sql("select make_country,
count(*) from cars group by make_country");
```

Finally, we fill this data into a `dataset` object. This `dataset` object is used later by the actual chart object:

```
    final DefaultCategoryDataset dataset =
        new DefaultCategoryDataset();
    List<Row> results = cdfByCountry.collectAsList();
    for (Row row : results) {
            dataset.addValue( row.getLong(1) , category ,
row.getString(0) );
    }
    return dataset;
```

2. **Create the chart component and fill it with the dataset object**: Using the JFreeChart API, we choose the type of chart we want to build and fill it with the `dataset` object we built earlier. We also specify the configuration of the chart such as the chart size, where we will display it (like at the center of the screen):

```
    JFreeChart barChart = ChartFactory.createBarChart(
            chartTitle,
        "Country",
        "Number of Cars",
            createDataset(),
        PlotOrientation.VERTICAL,
            true, true, false);

    ChartPanel chartPanel = new ChartPanel( barChart );
        chartPanel.setPreferredSize(new java.awt.Dimension(
        560 ,
        367 ) );
        setContentPane( chartPanel );
```

3. Finally, we just display the chart and center it on the screen. For the full code of this chapter, refer to the accompanied code in GitHub.

So, when would you use a bar chart?

Bar charts help to set up clear demarcations on your data and help to outline those in a pictorial form. When you plot a bar chart, you can easily figure out how data values compare to each other based on different criteria and it will help in understanding your underlying data better.

Data Visualization

Histograms

A Histogram is a special kind of bar chart. A histogram depicts some quantitative value on the *x* axis and frequency of that value on the *y* axis. The main feature of a histogram is that in a histogram, the *x* axes are grouped into bins and we treat each bin as a category. Thus, for a particular value, we take both the *x axis* bin and the frequency on the *y axis* into account.

Let's try to understand a histogram using the same `cars.json` dataset, which we used earlier. For the quantitative variable on the *x axis*, we will be using the number of cars grouped by each country and depict that on the *x axis*. The *Y axis* will denote the frequency of the number of counts, that is, the percentage or probability of countries with that amount of cars in the dataset. The diagram is as shown next:

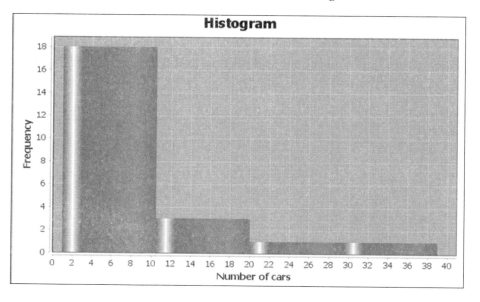

As you can see in the preceding chart, the maximum number of countries have a number of cars between 0 and 10 count. Next is the countries with cars between 10 and 20 count, and the remaining few between 20 and 30, and then 30 and 40.

When would you use a histogram?

For the purpose of data analytics histograms can be heavily used in the early data exploration phase and they give us a rough estimate of the density of our data. Thereby, from the preceding chart, you can figure out that our dataset has a lot of countries with only a few cars and there are only a few countries with lots of cars. If you check in the cars.json dataset you will see that the US and the UK have the maximum number of cars in this dataset. Sometimes, from a big data perspective, you might have to filter your data on this basis and segregate it, else your machine learning models might get trained with only one specific kind of data item that belongs to categories with a majority and hence they have to be normalized before being fed to predictive algorithms. We have explained this phenomenon in future chapters too.

There are particular shapes of histograms like symmetrical, right skewed (which we have in our preceding chart), left skewed, and so on. For more descriptions of histograms, please refer to the content on Wikipedia.

How to make histograms using JFreeChart?

In the following steps, we will see how to code a histogram using the JFreeChart library. We are using the same cars.json dataset here, which we previously used in the bar charts.

1. Create the HistogramDataset object to store data for the histogram: The approach here is also the same. First, we load the JSON file cars.json. Next, we register it as a dataframe temporary view and query it to pull the data and group the data by the make_country of cars. Next, we fill this data in a HistogramDataset object.

   ```
   Dataset<Row> df = sqlContext.read().format("json")
   .json("data/cars.json");
   df.createOrReplaceTempView("cars");
      Dataset<Row> dfc = sqlContext.sql("select make_country,count(*)
   from cars group by make_country");
      JavaRDD<DataItem> dataItems = dfc.javaRDD().map(s -> new
   DataItem(s.getString(0), new
   Double(s.getLong(1)).doubleValue())) ;
   List<DataItem> dataItemsClt = dataItems.collect();
      double[] values = new double[dataItemsClt.size()];
      for(int i = 0; i < values.length ; i++) {
          values[i] = dataItemsClt.get(i).getValue();
      }
   int binSize = values.length / 5;
      HistogramDataset dataset = new HistogramDataset();
              dataset.setType(HistogramType.FREQUENCY);
   ```

```
            dataset.addSeries("Histogram",values,binSize);
        return dataset;
```

2. Create the histogram chart object and provide the dataset to this chart object:

```
    private JFreeChart createChart(HistogramDataset dataset) {
    String plotTitle = "Histogram";
    String xaxis = "Number of cars";
    String yaxis = "Frequency";
    PlotOrientation orientation = PlotOrientation.VERTICAL;
    boolean show = false;
    boolean toolTips = false;
    boolean urIs = false;
    JFreeChart chart = ChartFactory.createHistogram( plotTitle,
xaxis, yaxis,dataset, orientation, show, toolTips,
urIs);
    int width = 500;
int height = 300;
        return chart;
    }
```

Line charts

These types of charts are useful in regression techniques as we will see later. It's a simple chart represented by a line that shows the changes in data either by time or some other value. Even Time Series charts are a type of line chart. Here is an example of a Time Series chart:

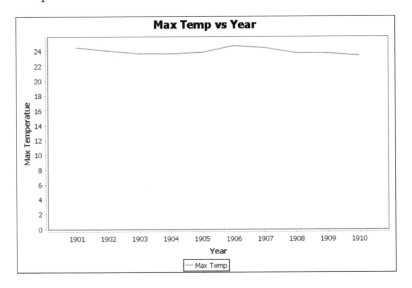

This line chart is a simple chart showing **Max Temp** versus **Year**, In this case, max temperatures are from **1901** to **1910**. The chart shows that the temperature did not change drastically within these 10 years.

To build this line chart, we have used the same `All India seasonal and annual min/max temperature series` dataset as explained in the preceding Time Series charts. For building the charts, the steps are again the same:

1. Loading the chart dataset and creating a JFreeChart-specific dataset.
    - We will create a similar `createDataset` method and return our `DefaultCategoryDataset` object

        ```
        private DefaultCategoryDataset createDataset() {
        DefaultCategoryDataset dataset = new DefaultCategoryDataset();
        ```

    - Next, we go on to build our boilerplate code for creating the `SparkSession`.
    - After building the `SparkSession`, load the `india_temp.json` dataset and register it as a temporary view.

        ```
        Dataset<Row> df = sqlContext.read().format("json").json("data/india_temp.json");
                df.createOrReplaceTempView("india_temp");
        ```

        Now, fire a query on this view to pull the first 10 records:
        ```
            Dataset<Row>  dfc = sqlContext.sql("select explode(data) from
        india_temp limit 10");
        ```

    - Collect the data and put it into the `dataset` object:

        ```
        List<Row> rows = dfc.collectAsList();
        for (Row row : rows) {
            List<String> dataList = row.getList(0);
            dataset.addValue(new Double(dataList.get(12).toString()),
         "Max Temp", dataList.get(0).toString()
        );
        }
        return dataset;
        ```

Data Visualization

- Code to load the chart: Create the `lineChart` object and fill it with the `dataset` object (see where the `createDataset()` method is invoked):

```
JFreeChart lineChart =
ChartFactory.createLineChart(chartTitle, "Year","Max
Temperatue",
createDataset(), PlotOrientation.VERTICAL, true, true,
false);
        lineChart.getPlot().setBackgroundPaint(Color.WHITE);

ChartPanel chartPanel = new ChartPanel(lineChart);
        chartPanel.setPreferredSize(new java.awt.
Dimension(560, 367));
```

So, when will you use Time Series charts? Whenever you have data points that refer to time, you can use this chart. As mentioned earlier, it will help you study the past data. Also, using the Time Series plot, you can estimate the future value.

Scatter plots

One of the most useful charts for data analysis are scatter plots. These charts are heavily used in data analysis, especially in clustering techniques, classification, and so on. In this chart, we pick up data points from the data and plot them as dots on a chart. In simple terms, scatter plots are just data points plotted on *x* and *y* axes as shown below. This helps us figure out where the data is more concentrated or in which direction the data is actually flowing.

This is very useful for showing trends, clusters, or patterns, for example, we can figure out which data points lie closer to each other. As an example, let's see a scatter plot next that shows the price of houses versus their living area.

As you can see from the graph, you will generally see that prices are going in the upward direction as the area is increasing. Of course, there are other parameters for the price to consider too; however, for the sake of this graph, we only used the living area. You can also see that there is a concentration of a lot of points in the 200K-500K price range and between 1000-1500 sqft area. Thereby, you can make a quick guess that a lot of people like to buy within this range of sqft area.

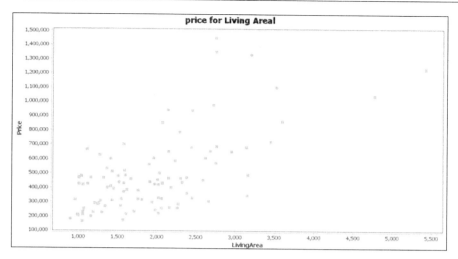

Making scatter plots with JFreeCharts is easy. As with other charts, we build the required dataset and chart component. We populate the chart component with the data we require and then plot the graph. For that, we follow few steps:

1. The code for creating the dataset component is shown next. The steps are simple, just load data via Spark from a text file. From the data, pull the living area and price and convert it to double. Now populate this data in a dataset object that is provided by JFreeChart.

    ◦ We will create the same `createDataSet` method but this time we return the `XYDataset` object using this code:

    ```
    private XYDataset createDataSet(String datasetFileName)
    ```

    ◦ Next, we have the boilerplate code to create Spark context and load the `textFile` for the dataset.

    ```
    JavaRDD<String> dataRows = sc.textFile(datasetFileName);
    ```

    ◦ Extract the living area and price and convert to double:

    ```
    JavaRDD<Double[]> dataRowArr =
    dataRows.map(new Function<String, Double[]>() {
            @Override
            public Double[] call(String line) throws Exception {
                    String[] strs = line.split(",");
                    Double[] arr = new Double[2];
                    arr[0] = Double.parseDouble(strs[5]);
                    arr[1] = Double.parseDouble(strs[2]);
                    return arr;
    ```

Data Visualization

```
            }
        }) ;
```

- Collect the data and fill it into the dataset object:

```
        List<Double[]> dataItems = dataRowArr.collect();
XYSeriesCollection dataset =new XYSeriesCollection() ;
    XYSeries series= new XYSeries("real estate item");
    for (Double[] darr : dataItems) {
            Double livingArea = darr[0];
            Double price = darr[1];
        series.add(livingArea, price);
}
        dataset.addSeries(series);
    return dataset;
```

2. Finally, create the chart component that you want to use from JFreeChart. In this case, it is a scatter plot as seen by the ChartFactory. createScatterPlot method:

```
  private JFreeChart createChart(XYDataset inputDataSet) {
     JFreeChart chart = ChartFactory.createScatterPlot("price for
Living Areal", "LivingArea", "Price", inputDataSet,
            PlotOrientation.VERTICAL, true, true, false);
     XYPlot plot = chart.getXYPlot();
        plot.getRenderer().setSeriesPaint(0, Color.green);
     return chart;
}
```

We are going to use the scatter plot heavily in our data analytics. This chart is also used along with other charts such as line charts to figure out how the data is properly segregated. As an example, refer to the following chart. Here the line chart is trying to divide the data points into categories. This can be used in classification to figure out the points into categories or labels based on whether they lie on the left-hand side of the line chart or right-hand side. You will see more of this in the coming chapters.

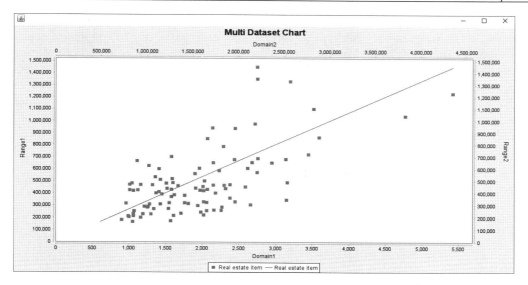

Now that you have seen scatter plots, let's answer a few simple questions:

When will you use a scatter plot?

Whenever you want to plot your data points on the x and y axes to figure out where the points are concentrated (that is, clustered), or which direction the patterns are flowing (that is, trends), or how data points are related to each other (that is, nearest neighbors), you can use a scatter plot. This plot will show you a cloud of data points that you can then use to figure out the correlations as mentioned earlier.

In the preceding example, you showed the graph only on the *x* and *y* axes, what if we have more features to take into account?

House prices are not just based on the living space area. There are many other features that can impact the price of a house. As an example, some of the additional features could be:

- Number of bathrooms
- Age of the house
- Zip code where it is located
- Condition of the house—whether it requires any additional maintenance work or not

So how would you plot a scatter plot when the number of features is high? When the number of features is very high, any feature size more than four or five is not easily comprehensible by human beings. This is because it is not easy for us to visualize 4- or 5-dimensional graphs. In real life use cases the features can be in the thousands, if not more. In such a scenario, a chart is represented by a figure called a **hyperplane**. A hyperplane is a plane or an area that you will try to fit within the n--dimensional space where n represents the number of features. So if you have two features, you will have a 3-dimensional graph with a simple hyperplane bifurcating your data points (represented by scatter points) as shown next:

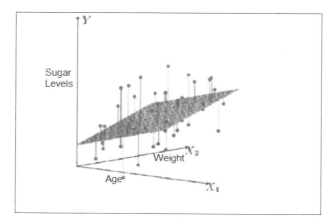

As you can see in the preceding 3D graph, the *x1* axis shows the **Age** of a person, the *x2* axis shows his **Weight**, and the *y* axis show the **Sugar Levels** of the person. The blue area depicts a hyperplane that is fitted to the data points for best case prediction. When the **Weight** and **Age** of a person is given what could be his **Sugar Levels**?

 We are not covering 3D charts in this chapter, but there are good 3D chart libraries both in Java or otherwise that you can use to plot graphs as shown earlier. Also, for even more dimensions, such as four or five dimensions, it's difficult to visualize it on a graph, but there are algorithms you can still fit in the hyperplane, though you won't be able to visualize the graph. In this case, you can plot the features with the data points in several separate graphs.

Box plots

Another very useful type of charts is box chart. Before looking into box charts, let's revise some simple mathematical concepts next. You can skip this page and directly go to the chart as well.

Suppose you have an array of numbers as shown here:

```
int[] numbersArr = { 5, 6, 8, 9, 2 };
```

Now, from this array, we have to find the following simple math stats:

- **Min**: This is just the minimum value from the array and as you can see it is 2
- **Max**: This is the maximum value from the array and this as you can see, is 9
- **Mean**: This is the mean value of the array elements. Mean is nothing but the average value. Hence in this case it is the sum of array elements divided by the number of elements in the array.

    ```
    (5 + 6 + 8 + 9 + 2) / 5 = 6
    ```

- **Median**: If we sort the preceding array in ascending order, the values would be:

    ```
    int[ ] numbersArr = ( 2, 5, 6, 8, 9 ),
    ```

    The value located at the middle of the dataset array depicts the median. As such, the median depicts a value in the array such that 50% of the values in the array are lesser than this and the other 50% of the values are greater than this.

    Thus, in our case the median is 6.

- **Lower quartile**: This depicts the value in the datasets such that 25% of values are lesser than this value. For ease of depiction, let's take a slightly bigger array for checking this:

    ```
    int[] numbersArr = { 11,5, 6, 8, 9, 2,7 };
    ```

    Now, if we sort this array, we will get the following:

    ```
    { 2,5, 6,7,8,9,11 };
    ```

Let's see at which index point in the array the first 25% of the values will lie. To figure this out, we find the index point where the first 25% of the values end. For this, we will use the length of the array and find 25% of that. We will later round off this value to the closest number and that value will be the index point in the array where 25% would occur. Therefore, the formula will be as follows:

*Number of elements in the array * (25 / 100) = 7 * (25/100) = 2*

So until the 2nd value of the array's 25% values are covered and since lower quartile refers to the value below which 25% values in the dataset are covered, we take the next value from our preceding sorted array, that is, the third value and hence it is 6. Thus, our lower quartile value is 6.

- **Upper quartile**: This depicts the value in the datasets such that 25% of values are greater than this value. Let's refer to the same array as earlier and find this. I have used the sorted array as follows:

    { 2,5,6,7,8,9,11 '} =>the value is '9'

The calculation of upper quartile that is shown above can be easily done using plain old Java code too. For this, we will build a simple quartile function in Java and in it we will consider the the lower percent as 25% and the highest as 75%. As soon as we enter the functions we will first check if the parameters passed are good or not, else we will throw an exception:

```
public static double quartile(double[] values, double lowerPercent) {
if (values == null || values.length == 0) {
   throw new IllegalArgumentException("The data array
   either is null or does not contain any data. ");
}
```

Next, we will now order the values and calculate our value by using the Math.round function and finally we will return the result:

```
double[] v = new double[values.length];
System.arraycopy(values, 0, v, 0, values.length);
Arrays.sort(v);
int n = (int) Math. round (v.length * lowerPercent / 100);
return v[n];
}
```

- **Outliers**: In statistics terms, an outlier is a value in the dataset that is very different from other values. It depicts that the data is not normally distributed and there are variations in the data and that you should be careful in applying analytic algorithms to that data, especially the algorithms that think that the data is normally aligned and has no abnormal values. Outlier points can therefore, indicate bad data or some errors in data.

Now that we have seen these simple math stats functions, let's dive into box charts.

So, what are box charts or box and whiskers charts?

It is a very convenient way in statistics to depict numerical data in terms of their quartiles, minimum, maximum, and the outliers. The whiskers or the lines that stretch out from the main chart rectangle boxes depict the values that stretch out beyond the upper and lower quartiles.

The following figure depicts one simple box chart:

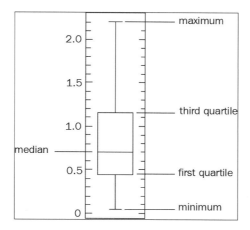

As you can see in the preceding figure, the rectangle is built with third and first quartile and the line within the rectangle depicts the **median** value. Also, the edges stretching out and the whiskers on them depict the minimum and maximum values. Now, the outliers on the chart can be shown as simple points as shown in the following figure:

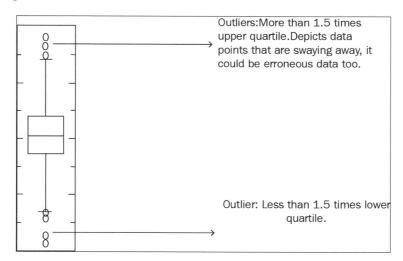

Box charts are very useful for studying numerical data and give a good overview of data distribution.

Let's try to understand the box charts using a realistic example. Suppose there is a fictitious website xyz.com that generates plenty of web traffic and you want to analyze the numerical hit counts in a generic way using some simple graph. The following table shows the stats of average hit count per day in a month:

Month	Hit count per day in the month
January	{25, 35, 45, 55, 60, 54, 34 ...}
February	{86, 90, 45, 55, 60, 54, 34 ...}
March	{54, 64, 89, 55, 60, 54, 34 ...}

These stats are stored in a CSV file that is pulled and parsed using Apache Spark. One row of a file is shown in the figure that follows.

If we now draw a box and whiskers chart on top of this data for the months between January and March, the chart would look like this:

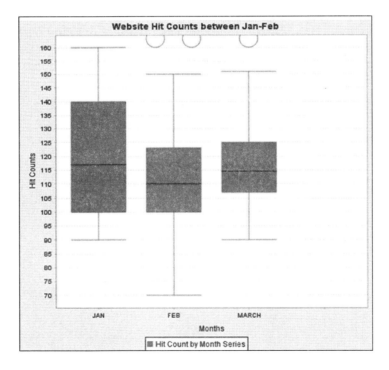

As you can see in the preceding chart, the outliers are marked by circular dots at the top of the chart. Also, as you can see for March, the maximum value was 150 and for February the minimum was 70.

JFreeChart has a very handy way to show all the stats such as mean, average, median, maximum, and minimum for a box chart. To see this, just mouse over on a particular chart and it will show you the details in a tooltip as shown in the following figure for January and so on.

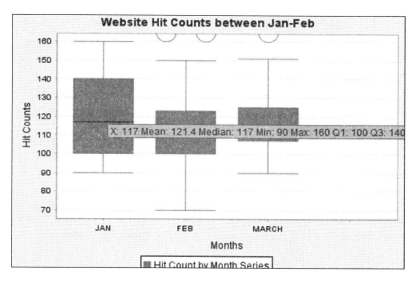

As you mouse over, the overlay window depicts the summary of stats such as min, max, first quartile (that is, Q1), and so on.

JFreeCharts have the API for box plots. As with other charts, we build the required dataset and chart component. We populate the chart component with the data we require and then plot the graph. The following are the steps we will use to build the chart for the dataset shown earlier:

1. The code for creating the dataset component is shown next. To create a box chart, we need a dataset of type `BoxAndWhiskerCategoryDataset`:
    - First we will create the `createSampleDataset` method that returns our specific dataset object:
      ```
      private BoxAndWhiskerCategoryDataset createSampleDataset() {
      ```

Data Visualization

- Next, we initialize our Spark `sqlContext` using our boilerplate code. Using the `sqlContext`, we now load the dataset file and extract tokens from it as strings (see the map method).

  ```
  sc.textFile("data/website_hitcounts.csv");
  JavaRDD<String[]> vals = rdd.map(s -> s.split(","));
  ```

- Next, we collect these values and fill our dataset object with the data:

  ```
  List<String[]> data = vals.collect();
  final DefaultBoxAndWhiskerCategoryDataset dataset =
  new DefaultBoxAndWhiskerCategoryDataset();

  for (String[] dataItem : data) {
  final List list = new ArrayList();
  for (int i = 1 ; i < dataItem.length ; i++) {
  list.add(Double.parseDouble(dataItem[i]));
  }
  dataset.add(list, "Hit Count by Month
  Series", dataItem[0]);
  }
  return dataset;
  ```

Finally, we create the chart component that you want to use from JFreeChart. In this case, it is a box chart as shown in the following code:

- First, we invoke `createSampleDataset` and store the result in a `BoxAndWhiskerCategoryDataset` variable.

  ```
  final BoxAndWhiskerCategoryDataset dataset =
  createSampleDataset();
  ```

- Next, for this specific chart, we have used `ItemRenderer`; this renderer object helps us customize this chart. So we can specify the details of the `ToolTip` renderer, or the specific custom *x* axis and *y* axis.

  ```
  final CategoryAxis xAxis = new CategoryAxis("Months");
  final NumberAxis yAxis = new NumberAxis("Hit Counts");

  yAxis.setAutoRangeIncludesZero(false);
  final BoxAndWhiskerRenderer renderer = new
  BoxAndWhiskerRenderer();
  renderer.setFillBox(true);
  renderer.setToolTipGenerator(new
  BoxAndWhiskerToolTipGenerator());
          renderer.setMeanVisible(false);
  ```

- We hook all the custom details of *x* axis, *y* axis, and the renderer in the category plot. Finally, we pass the `CategoryPlot` object to our `JFreeChart` object.

  ```
  final CategoryPlot plot =
  new CategoryPlot(dataset, xAxis, yAxis, renderer);
  final JFreeChart chart = new JFreeChart(
          "Website Hit Counts between Jan-Feb",
              new Font("SansSerif", Font.BOLD, 14),
  plot,
  true
  );
  final ChartPanel chartPanel = new ChartPanel(chart);
  ```

For full code, you can refer to our GitHub repository.

Advanced visualization technique

For advanced data visualization, commercial tools such as Tableau or FusionCharts can be used. These are very good in making dashboards and reports that can be used by businesses in their presentations or demos. In fact, for business needs, specifically for presentations or demos, we would urge the users to go with commercial tools such as Tableau or FusionCharts as they can be used to make very good reports and presentations. However, if you have specific advanced charting needs such as making three-dimensional charts or creating graphs or trees in Java, we can use advanced Java charting libraries such as Prefuse or VTK Graph toolkits.

Covering these advanced libraries in detail is beyond the scope of this book. Hence, we will only give specific brief outline on these libraries. Readers who are interested in these libraries can refer to their specific websites for more information.

Prefuse

This is an open source set of tools that is used for creating rich, interactive data visualizations in the Java programming language. It supports a rich set of features for data modeling, visualization, and interaction. It provides optimized data structures for tables, graphs, trees, and so on. The library is a little old but its code is available on GitHub at `https://github.com/prefuse/Prefuse`.

Data Visualization

IVTK Graph toolkit

As its wiki says, the **Visualization ToolKit (VTK)** is an open source, freely available software system for 3D computer graphics, image processing, and visualization used by thousands of researchers and developers around the world. It's an extensive visualization library that might be considered for your big data analytics work. You can check out this library at `http://www.vtk.org/download/`.

Other libraries

Apart from these, there are several other popular open source data visualization libraries written in other languages, for example, matplotlib in Python or D3.js in JavaScript. All these have different flavors and are useful to use on your big data analytics operations. D3.js is very famous and we would recommend you to use it if you are comfortable with JavaScript.

Summary

In this chapter, we covered six basic types of charts, namely, Time Series charts, bar charts, line charts, histograms, and scatter plots. These charts are extensively used in the data exploration phase to help us better understand our data. Visually understanding our data this way can help us easily figure out anomalies in our dataset and give us insights into our data that we can later put to use for making predictions on new data. Each chart can be used for specific needs such as:

- Time Series charts show us how our data changes with respect to time
- Bar charts show us the trends in our data and histograms help us find the density of our data
- Box charts help us find the minimum, maximum, median values in our numerical data, and also help us figure out the outlier points
- Scatter plots help us figure out patterns in our data or how our data points are concentrated

Java provides us with various open source libraries that we can put to use for making these charts. One such popular library is JFreeCharts that is heavily used in making charts using Java. We briefly covered an introduction of this library followed by making each type of chart using this library. We have followed a simple pattern in building these charts. We have loaded our datasets using Apache Spark and then used the JFreeChart library to build the charts. We believe that Java developers working on big data might find it easier to use a Java charting library initially for data exploration before they move on to more advanced charting solutions like D3.js and FusionCharts.

However, we are not restricting developers and they are free to use any framework of their choice for building the charts. Our aim for this chapter was to show you how different type of charts can be used in data exploration. In any analytics project that the readers might be involved in, depicted data using charts is a must, understanding the types of charts that can be used is as important as making the charts themselves. You can download the examples from our GitHub page and run these examples for practising these charts.

In the next chapter, we will study the basics of machine learning and learn how to handle a machine learning problem and the general approach in solving a machine learning problem.

4
Basics of Machine Learning

Any form of any analytical activity depends heavily on the presence of some clues or data. In today's world data is bountifully available. Due to the broad availability of various devices (such as mobile devices), IoT devices, or social network, the amount of data generated day by day is exploding. This data is not all waste; it can be used to make lots of deductions. For example, we can use this data to figure out what particular ad the user might click on next or what item the user might like to purchase along with the item they are already purchasing currently. This data can help us figure out a knowledge base that can directly impact the core business in many useful ways, hence it is very important.

This chapter is action-packed and we will try to cover a lot of ground while learning the basics. In this chapter, we will cover:

- Basic concepts of machine learning such as what machine learning is, how it is used, and different forms of machine learning
- We will look at some real-life examples where machine learning has been successfully used
- We will learn how to approach a machine learning problem and will see the steps involved in working on a typical machine learning problem
- We will learn to select features from data
- Finally, we will see how to run a typical machine learning model on a big data stack

What is machine learning?

Machine learning is a form of artificial intelligence where a computer program learns from the data it is fed or trained with. After learning from this data it internally builds a knowledge base of rules and based on this knowledge base it can later make predictions when it is fed new data. Machine learning is part AI, part data mining, and part statistics, but overall the criterion is to teach a machine to make new decisions based on past data it is trained with. So, for example, if we teach a machine some data regarding the inventory statistics of a store throughout the year then you might be able to tell things such as in which months the items sell more or which items sell more often. Also, it can tell the shop owner if they are selling one particular item more than other items; it can also show this to the customer so as to increase sales.

The concept of making new predictions is very important as we can now make predictions such as in which zone or area a marketing campaign should be launched first, or which segment of customers would be most interested in our new product; or if we increase our advertising budget by say a few percent then how much of an increase in sales we'll see. Such machine learning can directly impact business in a very positive way as it is totally coupled with real business use cases.

As we mentioned earlier, machine learning is part programming, part statistics, and part mathematics. Inference from data involves studying of existing data patterns, which is statistics, and doing this via a computer requires programming. Today machine learning is used in a lot of places to directly impact business. In fact, it is one of the hottest technologies to work in currently. Whether it is self-driving cars, suggestive searches, or customer segmentation based on their buying patterns, businesses are using more and more machine learning technology. Due to the tremendous interest in this technology in recent years a lot of new tools and frameworks have originated for machine learning and this is great for developers as now they have access to lots of resources for machine learning.

Before we delve into the code of machine learning algorithms let's look at some real-life examples of machine learning.

Real-life examples of machine learning

Machine learning usage is almost everywhere and it is growing as we read this book now. Here are some of the popular examples of machine learning usage where it has been running for years now:

- **Machine learning at Netflix**: Let's look at the following screenshot from Netflix. Look at the section **Because you watched Marco Polo**; it lists some movies or serials that might be of similar taste and would appeal to the user.

- How does Netflix figure this out? Basically, Netflix is keeping an eye on what the user is watching and based on that they try to figure out the user's likes and dislikes. Once a pattern of user likes is discovered a new set of movies are shown to them for viewing. In short, Netflix is analyzing user viewing patterns and then giving suggestions to the users:

- **Spam filter**: Spam filters that are used in our email account on a daily basis are a result of a good use of machine learning algorithms. Take a look at the following screenshot. Here we are showing two emails; one is spam and the other is not. The spam filter uses a special algorithm that uses some words (underlined in red) to figure out if an email is spam or not:

- **Hand writing detection on cheques submitted via ATMs**: This is a popular on cheque submitted via ATMs" usage implemented across a lot of banks already. Some of the best uses of machine learning are simple yet so powerful that they directly impact the day to day lives of many people. For example in this case when the user deposits a hand written cheque in an ATM machine, the ATM machine figures out the amount deposited by reading the handwritten amount from the cheque and the actual numbers in it.

For figuring the numbers from the cheque complex machine learning algorithms are used, but the whole process is very transparent and seamless for the end users. Look at the following figure, it shows a user depositing a cheque and receiving a receipt. The receipt shows the printed copy of the cheque, but before that it shows the amount that it read from the check:

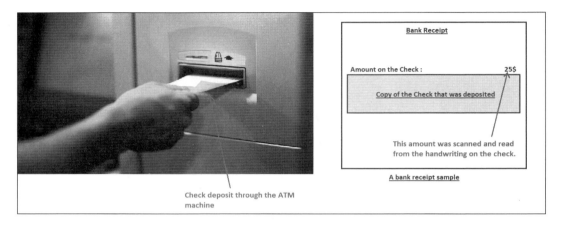

We will now go over the many types of machine learning algorithm.

Type of machine learning

There are three types of machine learning algorithm and within each type we have sub types. We will now go over the types of machine learning:

- **Supervised learning**: In this form, we have data that is labeled with results. The model is trained or fed with this prelabeled data. Based on this set of results the model internally builds its own knowledge and rule set. Using this knowledge that the model has learnt from existing data it can now classify a new set of data for the various labels. In simple terms this is what supervised learning is all about. And as the name says, the model is supervised with pre-existing data to make new predictions.

    Let's try to understand this with a simplistic example of a spam detection system. As seen in the following diagram, a set of words is used to form a dataset. The set of words is labeled as **GOOD** or **SPAM**. This dataset is fed to a model that builds its knowledge or rules based on this dataset. We call this training a **SPAM Detector Model**. Once the model is trained it can classify a new set of words as **GOOD** or **SPAM**:

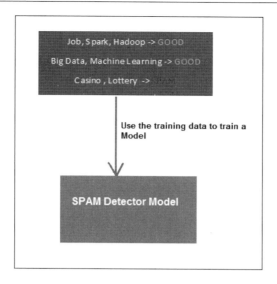

Now suppose a new email arrives, this new email can be broken into a set of words and those words can then be analyzed by this trained model to detect whether the email is spam or not, as shown in the following diagram:

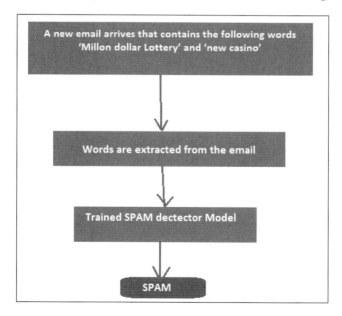

Supervised learning models are of different types. We will go over the different types of supervised learning now.

Here are a few types of supervised learning algorithm:

- **Classification**: In classification we have to predict to which category a particular set of data or attributes belong. So for example, if you take the `iris` dataset with three varieties of the same flower and teach your model to make prediction with that data, then later when you feed a new set of data to your model it will be able to classify the variety of flower based on the three varieties and the data that it was fed earlier.

- **Regression**: Regression is also a form of supervised learning as it is also based on existing data. But in the case of regression the value you are predicting is continuous in nature and not predicted from a set of categories as in the case of classification. Hence if you take the same `iris` dataset and build a model by feeding attributes such as sepal length, sepal width, and type of variety of flower, then later if you are given the type of flower and sepal length you might be able to predict a value for the continuous value sepal width.

- **Unsupervised Learning**: In this form we do not have any training data for the model. This is to say the model directly acts on the data, instead of making predictions about the data. We try to figure out the combinations or relationship between the data. Clustering is one form of unsupervised learning. With clustering we try to form groups between the data. These groups comprise data points that are similar to each other. For example, using clustering we can figure out islands of users that buy a product together, or figure out the area of epidemic in a region, and so on. Let's try to discuss clustering with an example. Let's look at the following graph, which shows the various datapoints that are collected for some disease outbreak in a region. All the datapoints are currently plotted in this two-dimensional graph as follows:

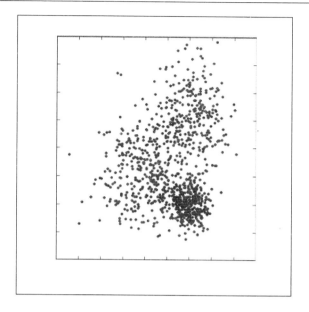

The preceding figure shows unclustered data in two dimensions for simplicity. Suppose the datapoints show the occurrence of disease in some region. Our task then is to group the data so that we can make some deductions from it. To group the data into different groups we run unsupervised clustering algorithms on these and try to figure out three specific groups within this region of disease outbreak. The groups so detected could be as follows:

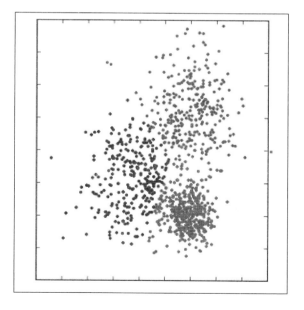

The preceding figure shows the clustering results after the algorithm is run on the raw dataset to figure out three groups. The three groups are shown in three different colors in the graph. Each could be some different form of deduction from the data, for example, the *red color* group might depict the region or area of severe form of the disease outbreak:

- **Semi supervised learning**: Semi supervised learning falls between supervised learning and unsupervised learning. In the cases of supervised learning we have a labeled dataset used for training and in unsupervised we have a completely unlabeled dataset that we use for finding the patterns. But in semi supervised learning we have a portion of the dataset that is labeled and another portion of the dataset that is completely non-labeled. Researchers have found that using a mixture of labeled and non-labeled data can sometimes produce great predictive results. In this book, however, we will be using mostly supervised and unsupervised learning algorithms.

Let's try to understand the concept of supervised and unsupervised learning algorithms using a simple case study now.

A small sample case study of supervised and unsupervised learning

Let's discuss a pollution detection system. Suppose we have an IoT device or any device that can capture the pollution levels or air quality levels and it is installed on various corners in the city. The role of this device is to pull the air quality levels at frequent intervals and store the results in files. These files can later be copied in a backend system where our analysis jobs can run on top of them. Also suppose we are collecting data for the following features: air quality levels, number of vehicles, location of the device, time of day, and the traffic congestion level.

Now suppose we have two analytic tasks on top of this data:

- Based on the features we have to predict if the number of vehicles is increased say two folds and time of day is in the afternoon when all the factories in the city are opened and producing toxic gases. What would be the air quality levels?

    This is clearly the case of supervised learning. We had a set of data initially where we had some features such as number of vehicles, air quality, time of day, and so on. We can feed this data into a machine learning model (model is a program or algorithm) and let it build its knowledge based on existing data. This knowledge is nothing but the relation between different data points based on which the program can make predictions in the future when the data changes.

So we can build and store a machine learning model. The next time if we need to predict air quality for a new set of features we can feed the set of features to the model and it can send us air quality levels.

This is pretty much what is going to happen in smart city projects that are on the increase recently.

- Task 2 is figuring out zones of high pollution levels within the city: This is an example of unsupervised learning where we do not have a result set and we need to just analyze the data and come up with a set of analytics. So in this case we figure out the air quality levels across different devices and figure out which quality levels appear to be similar and build zones based on these. This in other words is called clustering and is an example of unsupervised learning.

We have now seen the types of machine learning, let's now explore the typical steps in approaching a machine learning problem.

Steps for machine learning problems

A typical machine learning problem requires a set of steps to go through before we can start making an analysis using the machine learning models. The steps involved in the machine learning problems are:

1. **Understanding the problem**: Evaluating the problem at hand is a very important task. This step is basically used in understanding the problem and laying out the expectations from the outcome. This is basically where we analyze what we are trying to achieve from using machine learning for our problem.

2. **Collecting the data**: From the perspective of big data there are various ways of data collection. Data can be copied to HDFS or can be present in the NoSQL database HBase. Data that is stored in HDFS does not require to be frequently changed; however in HBase you can store data that can be changed and later reused. Data can be brought into Hadoop by variety of different ways, for example, you can bring in data in real time using messaging technologies such as Kafka or you can do a batch transfer of data via tools such as Flume or simple SFTP. We will be covering data ingestion in detail again in this chapter.

3. **Cleaning and munging the data**: When you work on a machine learning problem, it is best to first analyze the dataset at hand figure out what kind of data it is that you are going to train your model on. There could be some fields that might be null and have to be dealt with or some data that is not needed at all and can be filtered out. There could also be some data that is not in numeric format and needs to be converted into proper numeric format throughout to be fed to the models later. Analyzing and spending time on the data that is later fed to the models goes a long way towards making your models more qualitative in terms of predicted results.

    We covered this topic in *Chapter 2, First Steps in Data Analysis*, as well. Data is usually in raw format and contains some missing or bad values. Since our machine learning models take in mathematical numerical input, feeding raw or bad data to them would result in errors. Hence data has to be first converted into proper form or format before it can be consumed for data analysis. There are various ways of data cleaning and munging. Apache Spark has a handy API whereby you can create separate data cleaning Spark jobs and push the clean data back to HDFS for analysis.

4. **Exploring the data and breaking the data later into training and test data**: Initial data exploration is very important as it gives you the initial insights of the data and is a very important step in data analysis. Initial exploration involves plotting the scatter plots, or figuring out the number of rows in your dataset and you also do simple analysis as was shown in *Chapter 2, First Steps in Data Analysis*, using `spark-sql`. Exploring big data by writing simple `spark-sql` queries or plotting charts helps you better understand your data before you start running heavy analytics pieces on it.

    After you have run your initial data exploration break the data into two parts, one for training your models and the other part for testing on your models so as to check how well you have trained your model.

5. **Choosing, training, and storing models**: Choosing the right tool for the job is always of utmost importance. Think of the different models as different tools that you can use for different types of tasks; every tool or model has a unique set of features and drawbacks and it is your job as a data analyst to figure out the best fit for a particular use case. Apart from choosing a good model training the model with good data is very important too. Most of the models are mathematical and require data in number format, so make sure that data is in proper format (handle missing/nulls too) before feeding it to your models to train them. Once the models are trained they have to be stored (mostly external as in HDFS for big data models) so that they can be utilized later and applied on a new set of data.

> In real-world applications you would almost always use a group of models. Applying a set of tools for a particular task where the subtasks are handled by individual tools best suited for the job helps in getting a better job done, similar to applying a set of models (using techniques such as ensembling and so on almost always yields better results).

6. **Evaluating the results and optimizing the models**: Use the model that you have trained and stored in the previous step and apply it on the test data that you had kept earlier. Check how good your model is by using different evaluation techniques (such as root mean squared error and so on). Our aim is to bring down the error in our model predictions and reach a suitable level where our errors are minimized.

   To bring down the errors we will have to try different models, or change the input parameters of the models, or use a different set of features. In many cases it's a big trial and error process and this will consume a lot of time.

> You will see that beyond a certain point the error levels wont decrease further and as such that would be your maximum precision level. Also a good amount of data almost always beats the best-of-the-breed algorithms. The larger the data the better the knowledge set of the algorithms will become.

7. **Storing and using the results**: Once the models are good enough to be utilized, run them on your actual data now and store the results in external storage. Big data gives you various options to store your data. For example, you can store the results in Parquet format on HDFS, or as CSV or plain text in HDFS, or dump to the HBase data. The results stored in HDFS this way can be re-analyzed using `spark-sql` or using some real-time data analytics tools such as Impala.

We have seen the steps of running a typical machine learning problem, but still there are a few important questions unanswered and these questions are important for our understanding of big data and running machine learning algorithms on it. We will now explore some of those questions.

Choosing the machine learning model

Choosing the model depends upon the task we have at hand. If you have existing data available to train your models then you would be using one of the supervised learning algorithms either for classification or regression. If the end results are continuous in nature then you would use one of the regression algorithms and if the end results are one of some specific discrete values then you would use a classifier model. However, if there is no training data available and you still want to analyze your data then you would use some algorithms from the unsupervised learning models.

Within each type of machine learning technique, whether it's supervised learning or unsupervised learning, there are plenty of models to choose from. Before we look into the technique of choosing the model, let's look at two important concepts:

- **Training/test set**: It's a very useful practice and something that you almost always have to use. We would split the training data that we have into two separate sets. The first set we would use for training our model and the remaining set we would use to test our model. Since the remaining set will be totally new data for our model to test on, based on the error in predictions we can gauge the suitability of our model and improve upon it. Apache Spark provides us a handy method to split our dataset.

    First we load the data in a dataset object and then on the dataset object we invoke the `randomSplit` function providing the training and test set size (as shown by the ratio of `0.9` and `0.1`). `12345` is the seed value used to randomly split the data:

    ```
    Dataset<Row>data=
    spark.read().load("data/sample.txt");
    Dataset<Row>[]splits=
    data.randomSplit(newdouble[]{0.9,0.1},12345);
    ```

    As seen in the preceding code, the `splits` holds two values. The first value in the array is the training data and the other is the test data.

    However, even a plain training/test data holdout approach is prone to a common problem called overfitting. The reason is that we as a user of the model can still keep on tweaking it until it performs well on the test data. Thereby our model will perform nicely on this training and test data combination, but might again failout on new data. In order to train our model well we need to make our model touch as much data as possible for training, but still be good on new data. If we use too little data for training, our model might under fit, that is, it won't be nicely trained and would give bad predictions and if we use all the data it might overfit. A common approach to deal with this is to use cross validation.

- **Cross validation**: Cross validation uses the concept of training/test datasets also, but in this approach, we build multiple pairs of training and test dataset. K-fold cross validation is a common technique for cross validation and in this approach we build *k* pairs of training and test combinations. We train the model on the *k* training sets and test on each of the individual *k* test sets. Next we take the average of the error in predictions on each of the test sets and this is called the cross validation error. Our aim as model trainers is to reduce this cross validation error so that our model performs better.

For selecting a model you would first select it based on the type of problem at hand, that is, either a classification or regression or unsupervised learning algorithm. Once you have the algorithms you should train and test them with the cross validation approach and pick the model with the least amount of error on prediction results.

Let's now look at the types of features we can extract from our datasets.

What are the feature types that can be extracted from the datasets?

We will now list some of the main types of features that you will be dealing with on a day to day basis as a data analyst:

- **Categorical features**: These are discrete variables, that is, they are bound and have definite values. They are present as labels or strings to depict the outcome of a set of data points in a dataset row. For example in a dataset containing the health records of patients a feature like blood pressure might be represented in categorical form as high, low or normal. There are machine learning algorithms like decision trees and random forest that can consume categorical features as-is. However there are other algorithms like logistic regression that are purely mathematical and would require even the discrete categorical features to be converted to numeric format before training on them. For such algorithms we can convert categorical features to numerical format as shown:

    1. **Extracting features from categorical variables (continuous numbers)**: These are variables that are non-numeric. As most of the machine learning algorithms run on mathematical numeric numbers we need to convert the categorical non-numeric features to numeric features.

As an example, let's look at the following rows of data from some sample dataset that checks the health of a person based on some parameters or features and then tells if a person has a disease or not:

Sample data (to show examples of categorical data)
6,148,72,35,0,33.6,0.627,50, "has disease"
1,85,66,29,0,26.6,0.351,31, "no disease"
8,183,64,0,0,23.3,0.672,32, "has disease"

This is a classic example of a **binary classification dataset**. But as we know most of our binary classifier models (programs that can learn from existing data and build the ruleset), only work on mathematical data so we can't feed the last value, that is, "has disease" and "no disease" to our classifier models. Since "has disease" and "no disease" are the only two **categorical** values in this dataset, we can safely turn them into **numerical** values as "1" and "0" thereby the dataset would become:

Sample data (to show examples of categorical data)
6,148,72,35,0,33.6,0.627,50, 1
1,85,66,29,0,26.6,0.351,31, 0
8,183,64,0,0,23.3,0.672,32, 1

The previous example can now work well in this case of binary classifications.

2. **Extracting features from categorical variables (non-continuous numbers)**: In the previous point we showed you how to use continuous numbers 0 and 1 in a binary classification problem to replace categorical values. But does the use of continuous numbers (that is, numbers in order 1, 2, 3) work in all cases? The simple answer is no and in fact in some case continuous numbers can cause machine learning algorithms to produce erroneous results.

Not all categorical variables can have continuous numeric values as that might cause the models to treat them as numbers in order. Consider this example of a dataset row:

Dataset containing multiple categories of data points.
12,15,16,178,36,89, "New Jersey"
12,15,23,178,36,89, "Texas"
37,33,44,12,12,33, "New York"

Suppose the rows of data show few values based on which we classify whether the state referred here is New York, Texas, or New Jersey. Now if we just blindly put the values as 0 for New Jersey, 1 for Texas, and 2 for New York it won't be correct. The reason for this is that these are not continuous values; there is no relation between New Jersey, Texas, and New York. Hence we should replace them with non-continuous numbers that do not depict any kind of relationship like one number being greater than other (so it should have more emphasis), hence if New Jersey is say 14, Texas can be 50, and New York can be 6. None of these numbers show any kind of continuous relationship.

However, if you had categorical labels for movie ratings such as bad, good, and awesome, you could replace them with continuous numbers such as 1, 2, 3. Thus 3 being awesome has a value greater than the other two.

- **Numerical features**: These features are as real numbers or integers and can be extracted from raw data for analysis. Most machine learning models are dependent on numerical data. But even with numbers you might want to transform the numbers into a particular range to feed to your models.
- **Text features**: Features can also be extracted from plain text. Text can be in the form of comments on a topic, in the form of some reviews such as Amazon reviews for its products, or they can be as messages sent on Facebook, WhatsApp, or Twitter. This text is valuable as we can analyze this to figure out things such as current trends or the overall review of a product or movie, or do sentimental analysis of a set of text messages, and so on from text using **Natural Language Processing** (**NLP**) techniques.

Apart from these, features can be images and videos too. But these features have to be converted into different numbers using special techniques that are beyond the scope of this book.

We have seen the types of features; let's now look at the main methods of extracting the features.

How do you select the best features to train your models?

Feature selection is a very important task and it is tightly coupled with the predicted results. Especially in the case of big data analysis a dataset can contain thousands of features. We cannot use all the features as it can slow down the computation tremendously as well as yielding improper results. Besides this some of the features might be unnecessary or redundant. To overcome this problem various techniques on the feature selection side can be used to select a subset of features. Doing this would help us train our models better as well as reduce the problem of overfitting.

[Overfitting is the problem where our models are nicely fitted or trained on the training data. As such they work very well on instances on the training data for prediction results, but they work poorly on any new set of the data (that is, new data that the model has not seen before).]

Feature selection techniques are different from feature extraction as in feature selection we reduce the number of features we are using to train our models, but in feature extraction we are only interested in choosing features from our raw data. To depict the importance of features let's take a small example of a dataset shown as follows:

Age	Diabetic	Heart disease	Exercise	Smokes	Talks fluent English	Random variable
42	No	No	No	Yes	Yes	$#@
45	Yes	No	Yes	Yes	No	-
68	Yes	Yes	No	No	Yes	test

Now suppose somebody asks us to write a machine learning program that will teach a model to predict the age of people based on the features depicted in the preceding dataset. Look at the features shown — **Diabetic**, **Heart disease**, **Exercise**, and **Smokes** do make sense as they impact the age of a person. But what about **Talks fluent English** and **Random variable**. Random variable contains only garbage variables, not good to teach to the model and Talks fluent English is completely irrelevant. So we should discard these features and not use them to teach our model or else we will get bad results. This is what feature engineering is all about, choosing or building a proper set of features to teach your models with. It is a broad topic and entire books have been written on it. We will try to cover this in as much detail as possible as part of this book.

As part of feature selection there are three important techniques for feature selection and Apache Spark provides API methods for those techniques. The techniques that are used for feature selection are:

- **Filter methods**: As the name suggests we filter out the irrelevant features and choose only the relevant ones. There are many techniques by which we can only select a subset of the features we have and still get good predictions. There is a big advantage of choosing fewer features as it helps train our models faster, it helps in avoiding overfitting, and it might help in getting better results as with more features we might overfit and wrongly train our models. Simple statistical methods are used to correlate the features with the outcome variable and the features are selected based on that. Some of the filter methods are:

    - **Pearson coefficient**: Pearson correlation is a simple method used for understanding a relation of a feature with respect to its outcome or response variable. It measures linear correlation between two variables that are both continuous in nature (that is, numerical). The resulting value lies in *[-1;1]*, with *-1* meaning perfect negative correlation (as one variable increases, the other decreases), *+1* meaning perfect positive correlation, and 0 meaning no linear correlation between the two variables.

        Spark comes with some handy pure statistical functions built into its statistics package. The pearson coefficient function is also built inside it and using this function from the statistics package of Spark, you can apply it on RDD's of data. As an example, let's look at the following code:

        First we create two datasets `seriesX` and `seriesY` with some sample data:

        ```
        JavaDoubleRDDseriesX = jsc.parallelizeDoubles(
            Arrays.asList(1.0, 2.0, 3.0, 3.0, 5.0));
        JavaDoubleRDDseriesY = jsc.parallelizeDoubles(
            Arrays.asList(11.0, 22.0, 33.0, 33.0, 555.0));
        ```

Now we find the correlation between these two datasets and we find it using the `Statistics.corr` method from the statistics package of Apache Spark. Next we print out the result:

```
Double correlation = 
Statistics.corr(seriesX.srdd(), seriesY.srdd(), "pearson");

System.out.println("Correlation is: " + correlation);
```

```
Correlation is: 0.8500286768773001
```

- **Chi-square**: When both the feature and the response are categorical then the chi-square method for feature selection can be used. It is a statistical test applied to the groups of categorical features to evaluate the likelihood of correlation or association between them using their frequency distribution. Using this method is simple, we calculate the chi-square distribution between each feature and the response and figure out if the response is independent of the feature or not. If the response is related to the feature then we keep it or else we discard it. Spark ML comes with the chi-square feature selector built in. Let's try to understand it using an example.

    Suppose we have a dataset as follows (the column on the left shows the set of numerical features and the column on the right shows the response, this is just some sample data for this example):

Features (4 features per row)	Response
[0.0, 0.0, 18.0, 1.0]	1.0
[0.0, 1.0, 12.0, 0.0]	0.0
[1.0, 0.0, 15.0, 0.1]	0.0

    Now our task is to find out which among the four features is the top feature for predicting the outcome response (right column in the table).

    For this first create the dataset (here we are creating a dataset using sample data shown in the preceding table and providing a schema for that data; the full code is in our GitHub repository for this):

    ```
    Dataset<Row>df=spark.createDataFrame(data,schema);
    ```

Now create an instance of ChiSqSelector and set the number of top features that you want from it:

```
ChiSqSelectorselector=newChiSqSelector()
.setNumTopFeatures(1)
.setFeaturesCol("features")
.setLabelCol("clicked")
.setOutputCol("selectedFeatures");
```

Now apply this chi-square selector to the dataset we loaded earlier and extract the top features from it. Store the result in another dataset and finally print out the results from this dataset:

```
Dataset<Row>result=selector.fit(df).transform(df);

System.out.println("ChiSqSelector output with top "+selector.getNumTopFeatures()
+" features selected");
result.show();
```

The results will be printed as shown:

```
+---+------------------+-------+----------------+
| id|          features|clicked|selectedFeatures|
+---+------------------+-------+----------------+
|  7|[0.0,0.0,18.0,1.0]|    1.0|          [18.0]|
|  8|[0.0,1.0,12.0,0.0]|    0.0|          [12.0]|
|  9|[1.0,0.0,15.0,0.1]|    0.0|          [15.0]|
+---+------------------+-------+----------------+
```

> Note: Apache Spark comes bundled with other feature selection methods such as VectorSlicer and RFormula, please refer to the official Spark documentation for information on those.

- **Wrapper methods**: The concept of wrapper methods is simple. We pick a machine learning model and train it with a subset of features. Next, we record the error in our predictions and check how the errors change by choosing a different set of features (by removing or adding features from or to our original subset). We keep on doing this until we reach an optimum set of features .As such this method is very computationally expensive and it takes a long time to run and check. Correlation methods on the other hand are much faster, but they do not bring such good results as wrapper methods.

There are some common examples of wrapper method approaches. We will explain some of them now:

- **Forward Selection:** The concept is quite simple. We pick a machine learning model (it can be a decision tree) and train it with no features. We now check how the predictions are and the predictive results error rate is. Next, we keep on adding one feature at a time and keep on checking the error rate. If the feature improves the performance of a model, then we keep it; if not we remove it. Thus, it is an iterative and computationally expensive process, but it helps us build a good set of features.

> The feature set thus obtained might be tightly coupled with the model you have trained with. As such they might not be good for the other models. Also, this technique is prone to overfitting where by the models are good in predicting on training data, but predict badly on new test data.

- **Backward elimination**: This approach is just the opposite of forward selection. Here we pick a model and train it with all the features and next we keep on removing one feature at a time and observe the results. We remove the features that have no impact on model performance. This approach also suffers from the same problems as those of forward elimination.

Apart from these there are other approaches such as recursive feature elimination. We urge the readers to check on Wikipedia for more information on these.

- **Embedded methods**: As the name suggests these are methods that are inbuilt or embedded within the machine learning algorithm itself. Thus, all features are fed to the model and models containing the embedded method will pick the best subset of features by themselves. These embedded methods combine the qualities of wrapper as well as filter methods both. Some of the popular examples of embedded methods are lasso and ridge regression. These methods lasso and ridge regression have inbuilt penalization functions to reduce overfitting.

> Note: Regression in statistics refers to the technique of adding additional information to improve the performance of a model.

We have seen now how to select the models and how to select its features. Let's now see how we can run a complete machine learning algorithm on big data.

How do you run machine learning analytics on big data?

The real usage of machine learning comes mainly in the form of big data. Over a period of time you will realize that more data beats best-of-the-breed algorithms. More data means more criteria's and more knowledge that can be fed to the models and they will then produce better results. Most of the companies that are heavily using some form of analytics for business decisions are now also using or getting into big data. The reasons are:

- In some cases, it might be completely impossible to run the analytics on traditional datasets, for example, consider the case of storing video's and images. Relational databases have a capacity beyond which storing data in them just does not makes any sense and they won't scale beyond a certain point. Hadoop is especially suitable for storing this kind of complex data such as videos and images.

  Now what if your task is to analyze the videos and images and extract out any vulgar content such as porn images or videos. There are different forms of machine learning like deep learning that you can use to classify images into porn and filter them out. To run analytics jobs of this scale where millions of images and videos are involved is specially suited for running on big data with cluster computing frameworks such as Apache Spark.

- Hadoop is open source, so it is easily available and there are lots of vendors that sell support for big data stacks such as Hortonworks, Cloudera, MapR, and others.

- Network cost is low due to parallel computing and data locality.

- Parallel jobs reduce the total execution time of computations considerably. Earlier, a lot of computations for example computations on spatial data or genomic data used to take days when processed through sequential batch jobs but now can be run much faster as they can be processed in parallel now. Easy rollback and failover support from different parallel jobs.

We have seen the advantages of running a machine learning problem on big data. Let's now see how we can run a typical machine learning problem on big data. We will try to cover the individual steps in detail.

Getting and preparing data in Hadoop

Hadoop is a very open ecosystem as such data can be fetched into it from various datasources and by various methods. It has a plethora of products now that can be used for batch analysis of data as well as real-time analysis of the data. There are plenty of sources from which data can be collected and pushed into Hadoop for storage, for example, all the images, videos, tweets, logs, or data from existing apps can be dumped into Hadoop. So Hadoop becomes a big storage engine for regular apps data as well as social media apps data and here all this data can then be analyzed to provide useful business intelligence.

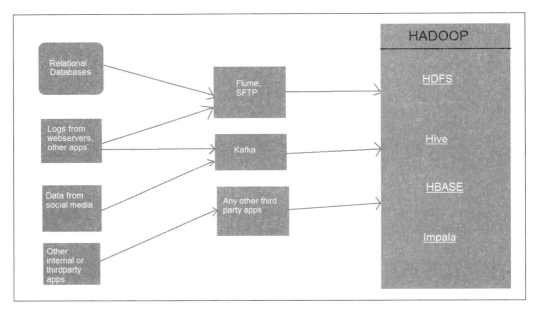

As seen in the preceding figure, there are various sources from which data can be pulled into Hadoop. Also various technologies such as Flume, SFTP, Kafka, and so on can be used to transfer data from sources into Hadoop. We will cover some of these sources and data transfer techniques now:

- **Flume**: Using Flume you can directly copy the `log` files or other data files to HDFS.
- **FTP**: You can FTP the files to some location on the shell (Linux shell, and so on) and can later push this file to HDFS.

> Note: If you are playing around with a large dataset and planning to run data analytics on it you can explore the option of simply FTPing the file and later pushing it to HDFS manually.

- **Kafka**: You can publish the data to a Kafka topic and this data will be pushed by Kafka to HDFS. This is more of a real-time method that is generally used in event driven systems, for example, if you want to collect tweets from Twitter and every few seconds or at intervals you want to publish them to HDFS then Kafka is great for such usage.

    You can use Apache Spark's Spark Streaming API to read the data from the Kafka topics and store them on HDFS.

- **HBase**: This is the default database that ships along with most Hadoop bundles. Data can be inserted into this database and then can directly be consumed for analysis. This is the main database that comes into action when you are sending a message to your friend on Facebook.

- **Hive**: This is a batch solution. So if you push data into it it stores the data in a Hive datawarehouse that is kept inside HDFS. Hive has an SQL engine inbuilt so you can fire SQL queries for analysis. Hive is slow as it is batch and internally for queries it fires MapReduce jobs, but it is still used in many places due to its good support for SQL.

- **Impala**: This is a more real-time option This product was built by Cloudera. It is very fast and is an excellent product for quick real-time data analytics on big data.

> For running the examples in this book we would suggest you copy the files to HDFS.
>
> For this you need to copy the file to the Linux or Unix shell first and later push them to HDFS using the following command:
> `Hdfsdfs -put <FILE_NAME><HDFS_LOCATION>`
> This will pull the file from the shell and push it to the HDFS location.

We have seen how we bring data into Hadoop, let's now see how we prepare this data in Hadoop.

Preparing the data

Data might be completely raw when it is first brought in Hadoop. As we discussed earlier data has to be cleaned and munged thoroughly before it can be consumed by machine learning models. In Hadoop, and generally in data warehousing, raw data is initially copied to a staging area (temporary area) and then some Spark jobs or MapReduce jobs can be run on them to clean the data and republish the data into HDFS.

Formatting the data

From the perspective of the data 'form' we must compress the data before it can be dumped into HDFS. There is no restriction though on the type of data that can be stored in Hadoop. You can store data in text, JSON, binary, or any other format. However, for applications in production we normally prefer data in some compressed format for example Avro or Parquet format. The reason being that the compressed data will take less space and it would need less network bandwidth for data transfer. We would recommend that the users use formats of data that are well suited for Hadoop itself, for example, Avro, Parquet, or SequenceFile. Apart from compression there is one other major advantage of using these specific compression formats on Hadoop and that is they support partitioning, thus an Avro file can be easily split and spread across multiple machines without corrupting the data it holds. Please refer to Hadoop documentation for more details on this.

> Note: If you are going to run lot of aggregation analytical tasks such as aggregating on specific columns then it's better to use Parquet format as it is well equipped for cases where we require fewer columns for data analysis.

Storing the data

After the data is properly cleaned and formatted it can be stored in Hadoop, as was shown in the previous figure, either directly in HDFS (in Parquet, Avro, or any other format) or it can be passed to HBase or Hive.

While storing the data in Hadoop you might have to check how you partition the data as this is a very important step and it guides the performance of an application running on top of this data. For example, if you are storing data dumps from various dates then you can partition the data on date, that is, you can store it as:

```
<HDFS_DIR>/data_ingestion_date=<DATE_Value>
```

Here `HDFS_DIR` is the director containing the HDFS folder that begins with `data_ingestion_date=` and after this equals sign we have the actual date when the data dump was taken.

This helps to partition the data by date and the advantage of partitioning is simple yet very powerful. So suppose you need the data only for the last two days and then you just need to check the folders for the last two days in HDFS and you can simply ignore the remaining folders. This would mean that if you had been taking a dump of your data every day for few months or years and putting each days data in its specific folder (that could have a 'days date' as its name) in HDFS, you would simply ignore that old data and only take the latest data for last two days. This would highly expedite the queries or jobs run on data as the amount of data analyzed is reduced now.

Data partitioning on Hadoop is a very important topic and please refer to the Hadoop official documentation for more information on it.

In this section, we have covered how we can fetch data into Hadoop and how we can prepare it so that is ready to train our machine learning models. In the next section, we will see how machine learning models can be trained on big data.

Training and storing models on big data

Most data scientists from the Python world are used to working on libraries such as scikit-learn and these are mainly single node libraries that do not work well on big data. Big data requires specially built machine learning algorithms that are designed to work on a cluster of multiple machines. Apache Spark ships with a lot of machine learning algorithms bundled in its machine learning library. These algorithms are specially designed to work on a cluster of distributed machines as such they are best suited for big data analysis. Since Apache Spark is a relatively new library not all the algorithms are available. So if you want to use a machine learning algorithm from some other library then you need to use the `jar` file in which it is bundled and put it in the classpath of all the datanodes that are running your Spark jobs. This way on each datanode where your Spark job runs the specific JAR would be available.

These are the steps to build and train a model on big data:

- **Choose the model**: If the model is present in the Spark implementation choose it first and build your Spark program on top of it. If the model is not present in the default Spark machine learning library then use the library you want to use and write your Java program on top of it. Next you can invoke your Java program using a Spark job, you just need to provide this jar in the classpath for the Spark job.

- **Train the model**: Train the model on the clean data that you prepared earlier and test it on the test data. After making different iterations of testing your model when your model is good you can store it.
- **Store the model**: After the model is trained it should be kept in external storage so that later it can be repicked and used for testing. This way you won't have to train a new model again when you want to do some prediction results again.

> In the case of big data models, storage is a very important operation as the amount of data involved is huge; you cannot retrain the model again on the same data. So it's better to train and keep a good performing model in external storage and reuse it when needed.

In Spark ML all the machine learning algorithms have the `save` methods using which a trained or fitted model can be exported to external storage and later reused. Let's see an example of this API:

Here we have a Logistic Regression Model:

```
LogisticRegressionlr=newLogisticRegression();
```

After creating the model, we train it on some `training` data:

```
LogisticRegressionModelmodel1=lr.fit(training);
```

Once the model is ready and trained, now you can store it to its HDFS location (for example, as on `hdfs://testapp/modelstore/temp`):

```
model1.write().overwrite().save("hdfs://testapp/modelstore/temp");
```

Now once the model is saved in this external location on HDFS, you can always deserialize it and bring it back into an object of the same model type by recreating it from external storage:

```
LogisticRegressionModelmodelFromExternalStore =
LogisticRegressionModel.load("hdfs://testapp/modelstore/temp")
```

As seen here, each Spark ML model contains a handy function `load` using which the model can be loaded from external storage like HDFS. Apache Spark has a very extensive machine learning API that is suitable for feature extraction from raw data, feature selection, machine learning algorithms, and some utility functions. We will learn more about this API in the next section.

Apache Spark machine learning API

Spark ML is Apache Sparks library for machine learning analysis. This library contains machine learning algorithms that are pre-designed to run on a cluster of distributed machines. This feature is something that is not available on the other popular machine learning libraries such as scikit-learn as such these are single node libraries. To run these third-party single node libraries via Spark you will have to ship their code on each individual machine that is running a Spark job. You can do this via the `spark-submit` job.

Apart from this Spark machine learning algorithms are massively scalable and are much easier to write and maintain as compared to older versions of machine learning algorithms built on top of map reduce. Algorithms built on top of MapReduce were slower, much more complicated in terms of code, and were hard to debug and maintain.

We will go over the specific details of the Spark ML API now.

The new Spark ML API

The initial machine learning algorithms of the Apache Spark (MLlib API) were highly centered around the RDD API. But over a period of time they started concentrating on the DataFrame API, hence in this book all machine learning algorithms are based on machine learning algorithms run using the DataFrame piece. Those people coming from a scikit background will find the DataFrame API much similar to the one they use. Overall the DataFrame machine learning API is very simpler to use and maintain, it has handy tools for feature transformations and extractions.

> Note: As of Spark 3.0 the MLlib RDD API will be fully deprecated hence we encourage the users to concentrate on using the DataFrame machine learning API wherever they can.

At a higher level, the Spark ML API contains the following tools:

- **Machine learning algorithms:** Spark contains the implementation for some popular machine learning algorithms, for example, for Logistic Regression, Naive Bayes, clustering, and so on. The API is very developer friendly and is very easy to use. If you have used libraries such as scikit-learn you will find lots of similarity in terms of usage. The following is an example of sample code where we load the dataset and apply a k-means clustering algorithm:

Basics of Machine Learning

We are going to first load our dataset from a file and store it in a dataset variable `dataset`:

```
Dataset<Row> dataset = spark.read().format("libsvm").load("data/mllib/sample_kmeans_data.txt");
```

Next we create a `KMeans` algorithm instance and apply the algorithm to the dataset:

```
KMeanskmeans = new KMeans().setK(2).setSeed(1L);
KMeansModel model = kmeans.fit(dataset);
```

Our task here was not to explain the code or algorithm to you, but we just wanted to show how easy it is to use the Spark ML code. We just need to load the dataset in the proper format and instantiate an algorithm and apply to it. We can also pass custom parameters to our algorithm.

- **Features handling tools**: Feature engineering is one of the most important fields within the machine learning arena. It's not the number of features, but the quality of features that you feed to your machine learning algorithm that directly affects the outcome of the model prediction. Spark ships with feature handling tools that help in doing feature extraction, transformation, and selection. It's a very handy set of prebuilt feature tools and we discussed them in detail in an earlier section in this chapter.

- **Model selection and tuning tools:** These are methods that are specifically used in choosing a model as well as tuning its performance. Cross validation is built into the Spark ML package and can be used to best train a model and hone its performance.

- **Utility methods:** The API contains some utility methods to do some basic statistics as well as some other basic utility methods.

Some of the useful utility methods are:

Method	Details
`Statistics.colStats()`	This method returns an instance of `MultivariateStatisticalSummary`, which contains the column-wise max, min, mean, variance, and number of nonzeros, as well as the total count. This method is applied on an instance of RDD[Vector] and is part of the `Statistics` package with Apache Spark MLlib.
`Statistics.corr()`	We covered this in a previous section too. Spark provides methods to calculate pearson as well as spearmans coefficient and these methods are present in the `Statistics` package only.

Method	Details
RandomRDDs	These contain factory methods for random data generation, very handy in prototyping and testing a model with some randomly generated data.

Summary

This chapter was action packed on machine learning and its various concepts. We covered a lot of theoretical ground in this chapter by learning what machine learning is, some important real-life use cases, types of machine learning, and the important concepts of machine learning such as how we extract and select features, training our models, selecting our models, and tuning them for performance by using techniques such as training/test set and cross validation. We also learnt how we can run our machine learning models specifically on big data and what Spark has to offer on the machine learning side in terms of an API.

In the next chapter, we will dive into actual machine learning algorithms and we will learn a simple yet powerful and popular linear regression algorithm. We will understand it by using an example case study. After studying linear regression we will study another algorithm logistic regression and we will also try to learn it by using a sample case study.

5
Regression on Big Data

Regression is a form of machine learning where we try to predict a continuous value based on some variables. It is a form of supervised learning where a model is taught using some features from existing data. From the existing data the regression model then builds its knowledge base. Based on this knowledge base the model can later make predictions for outcomes on new data.

Continuous values are numerical or quantitative values that have to be predicted and are not from an existing set of labels or categories. There are lots of examples of regression where it is heavily used on a daily basis and in many cases it has a direct business impact. Some of the use cases where regression can be used are the following:

- To estimate the price of a product based on some criteria or variables
- For demand forecasting, so you can predict the amount of sales of a product based on certain features such as amount spent on advertising, and so on
- To estimate the hit count of an e-commerce website
- To predict the value of a house based on features such as number of rooms, living area, and so on

As you can see in the preceding cases, all the values predicted are continuous numerical values and even though the model is trained with data from existing values the outcome value is quantitative and does not lie from a predefined set of values.

In this chapter, we will cover:

- The basics of regression, including what regression is and how it is used in machine learning
- We will learn about linear regression and see a real-world example of how we can predict housing prices using linear regression

- We will improve our linear regression model by using some standard techniques
- We will briefly introduce other forms of regression techniques and explain their benefits
- After covering linear regression to predict continuous values we will learn a popular machine learning approach logistic regression
- Using logistic regression we will study a real-world example of detecting heart disease in patients using a UCI dataset
- We will improve our model by using standard approaches
- Finally, we will also see the benefits of using logistic regression and in which places it can be used

Linear regression

As we mentioned earlier, regression is a technique for predicting continuous values based on certain inputs or variables. With linear regression, we try to learn from data that can fit into a straight linear line. For example, we can try to predict the amount in sales of a product based on variables such as amount spent on advertising, number of hits received on the e-commerce website, price of the product, and percentage offered in terms of sale price. To explain linear regression let's use a simple example using a sample fictitious data of the count of likes on a Facebook post versus the number of times it was shared, as shown in the following table:

Count of likes	Number of times shared
100	10
200	20
300	30
400	40
500	50
600	60

Let's try to plot this data on a line chart:

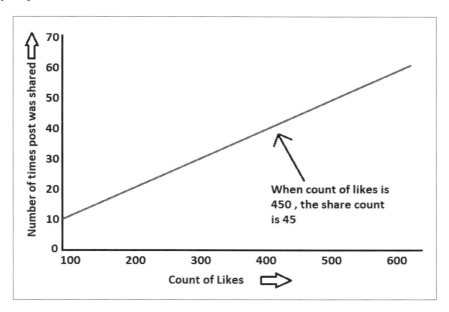

As you can see in the preceding figure, the data points are linear, which means that they linearly go up. In other words, an independent variable **count of likes,** when changed causes the value of a dependent variable that is **number of shares** to change. Thus, if we know a future value of the count of likes we can predict the value of the number of shares based on historical data that can be fed to a mathematical function. This linear mathematical function can be created based on the historical data. Now suppose I ask you to predict the value of shares count given that a particular Facebook post's likes count was 450. As seen by the arrow in the chart you can predict this value from the preceding line chart. The corresponding shares count comes to 45.

The preceding diagram depicts a very simple form of **simple linear regression**. Let's now dig a little deeper into simple linear regression.

What is simple linear regression?

Simple linear regression is a simple form of regression where we try to predict the value of one dependent variable based on changes to another variable. If we have a dataset with the value of the variables (x) and the labels (y) then simple linear regression can be represented by a mathematical formula:

$$y = ax + b$$

Using this formula, we try to predict the value of y when we get a new value of x.

For the purpose of our dataset, we will plot our dataset into a scatter plot, as shown in the following figure. All the data points will show up on the scatter plot. Then we will try to draw a simple line through the data points on the chart and try to find a best fit line. We will later use this line to predict the value of a datapoint given the variable x:

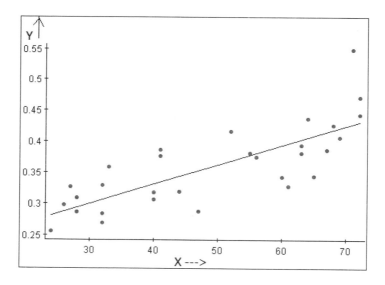

The preceding graph shows the line through the data and this line depicts the formula $y=ax+b$.

If we want to use the line to make predictions on future values, we need to find the values of the parameters a and b.

How do you find the values of a and b in the preceding equation ($y = ax + b$)?

The common approach for finding *a* and *b* is by making several lines through the data (scatter plot), as shown in the following figure. From the data points we now calculate the distance to our best fit line.

We have depicted this in the following figure using the lines between the data points and our prediction line.

Essentially, this is just the difference between the predicted value and the actual value. To compute the total error we sum up this error value for all our predicted data points. There is a problem however: some values are positive and some are negative. Due to this when you try to add the negative and positive numbers, the negative value would be subtracted from the positive one and this would reduce the absolute overall value of the error. To fix this issue, we take the squared of the error for each predicted and actual value difference, so this will always result in a positive number. Later, we sum up all the errors to find the net error. Taking the mean (average) of this net error would return the mean squared error, as shown in the following figure:

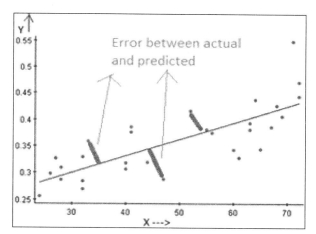

The preceding graph just shows one line and for the predicted values on that line we calculate the error (squared of error) and sum up the total errors. We do this again by creating a new line through our dataset and again, calculating the total error. We will keep on repeating this until we find our best fit line with minimal error (or after which point the error value does not change much).

 If we take the squared root of mean squared error we get root mean squared error and this value is quite popular in our regression calculations.

Up until now we have discussed only one feature, what if the number of features is more than one?

In this case, to best represent our data instead of a single line we will have something called a **hyperplane**. Let's see an example of a hyperplane, as shown in the following figure:

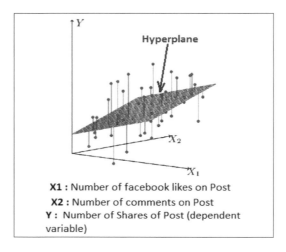

As you can see in the preceding chart, both the number of comments (**X2** axis) and number of Facebook likes (**X1** axis) are responsible for the quantity of share count (**Y** axis). In this case instead of the best fit line representing the data we have a best fit hyperplane (three-dimensional in this specific case) representing the data. When the number of input variables (that is, **X1** and **X2**) is more than one we call this linear regression **multiple linear regression**. We have seen that linear regression has been used for data points that are spread out linearly and can be depicted using a best fit line. However, what if the data points are like in the following figures?

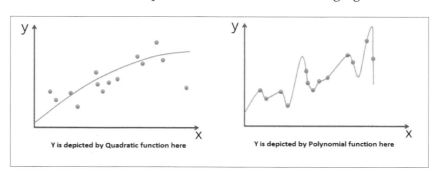

As seen in the preceding figures, we have a curvial relationship and a polynomial relationship. Both are still depictable by a linear regression model. In this case, the mathematical function depicting the outcome will change. If you see the polynomial relationship it fits the data nicely, but that does not mean it will produce great predictions simply because it fits nicely on training data, how this would perform on test data (that is, a totally new piece of data) is not known yet. The curvial relationship shown in the first figure is represented by the following mathematical formula:

$$Y = a + bx + cx2$$

(In this case, we need to find the parameters *a*, *b*, and *c*, we can use the same mean squared error principle as we used previously)We have seen some concepts of linear regression and now let's see its uses.

Where is linear regression used?

Linear regression has a number of practical uses. Some of the uses are listed here:

- Linear regression can be used in businesses to make estimates of product prices.
- Linear regression as it is depicted by linear graphs can be used to depict trends of items, for example, trending sales of an item over a period of time.
- It can be used in sales forecasting, for example, if the price of a product is reduced by some percentage linear regression can be used to forecast the amount in sales.
- In the field of finance linear regression is used in risk analysis.
- Linear regression can be used to predict the healthcare cost involved for individuals based on some variables. This information is very useful for health insurance companies as they can predict earlier how much cost might be involved in giving insurance to a person.

So much for the theory, let's dig deep into a real-life case study of linear regression. To run our samples we will use the Java code on the Apache Spark MLlib library. We will be showing an example of multiple linear regression and for simple linear regression we leave it as a task for the readers of this book, please check out the code on the main Apache Spark MLlib page and it has great working examples on it. Also note simple linear regression is inherently so simple that you can directly write a program on it and run it or use a `SimpleLinearRegression` class from the Apache Commons - project.

We are using Spark only to show the users how we can run these algorithms in parallel on big data, other programs such as `SimpleLinearRegression` from Apache commons are suited for running on smaller datasets. We are using the latest Apache Spark MLlib API using the `DataSet` object and not the RDD-based MLlib API. We encourage users to do the same too as the RDD MLlib API will be deprecated in Apache Spark 3.

Predicting house prices using linear regression

This is a sample case study where we will train our linear regression model using a training dataset. Once the model is trained, we will feed new test data to it to predict the house price for us based on the new data input. For this case study we will be using a dataset from http://www.kaggle.com. The steps that we will use in running our full algorithm of regression are as follows:

1. Collecting the data from the data source: In our case, since it is a file we can simple copy it to HDFS since it is a big data project and Apache Spark expects the file to be on HDFS or another big data filesystem to mimic a real-world application.

> The data in HDFS can be in any format such as plain text, CSV, Avro, or Parquet. Hadoop supports a variety of formats. Please refer to *Chapter 1, Big Data Analytics with Java* where we briefly covered a few of the popular formats.

2. Cleaning and munging the data.
3. Exploring the data for initial insights.
4. Building and training the regression model and storing the model to external storage on HDFS.
5. Reloading the model from HDFS and running predictive analytics for the price of house.
6. Evaluating our results.

We will now look into the full analysis of this sample application.

Dataset

The dataset contains house sale prices for King County, which includes Seattle. It includes homes sold between May 2014 and May 2015. This dataset has been used from `kaggle.com` and it can be downloaded from this link: `https://www.kaggle.com/harlfoxem/housesalesprediction`. Kaggle.com is a famous site that hosts data science competitions for developers and has plenty of datasets for learning analytics. This King County house sale dataset is a great dataset for evaluating regression models. Let's try to run our regression algorithm on this dataset. There are many features in the dataset, but the main features that we will be using are shown here:

Feature name	Description
Bed_rooms	Number of bedrooms
Bath_rooms	Number of bathrooms
Sqft_Living	Living area in squared feet
Sqft_lot	Lot area in squared feet
Price	Price of the house

Data cleaning and munging

Before we use any dataset for our machine learning tasks we have to make sure the data is in proper format. Most of the algorithms that we use rely on using mathematical data. Hence, data that is in string format as labels and so on needs to be converted into proper mathematical format before it can be fed to the machine learning algorithm.

Our first step for data cleaning is to load the data and visualize the first few lines and see if some garbage data is there or some missing values that you can visualize. Fortunately for us this dataset is pretty clean and it has no missing values. Also all the data points in it are numbers. Let's explore our data a little bit before we finally run regression on it.

Exploring the dataset

Before undergoing any machine learning task we must analyze the dataset first. In the event of some missed features or bad data we must clean the features data first before using it for training the models. Before we run any analytics job on our data we need to build our Spark configuration and `Spark session` object as shown in the following code:

```
SparkConf c = new SparkConf().setMaster("local");

SparkSession session = SparkSession
```

```
.builder()
.config(c)
.appName("chp_5")
.getOrCreate();
```

Once the `SparkSession` is built, let's run some simple analytics on data using Spark SQL as follows:

- **Number of rows in this dataset**: Using the `SparkSession` object load the dataset using the `csv` method as it is in CSV format. Next invoke `count()` and find the total rows in this dataset:

    ```
    Dataset<Row>data = Spark.read().CSV("data/kc_house_data.CSV");
    System.out.println("Number of Rows -->" + data.count());
    ```

    And the output is as follows:

    ```
    17/03/03 01:17:32 INFO DAGScheduler: Job 1 finished: count at HousingDataExplore.
    17/03/03 01:17:32 INFO CodeGenerator: Code generated in 31.059641 ms
    Number of Rows --> 21614
    17/03/03 01:17:32 INFO SparkContext: Invoking stop() from shutdown hook
    17/03/03 01:17:32 INFO SparkUI: Stopped Spark web UI at http://192.168.1.6:4040
    ```

- **Average price of houses per zip code sorted by highest on top**: For this we will register the dataset generated in the previous step as a temporary view named houses and then we will fire a simple group by query on this view. We will group the columns by zip code and find the average price per zip code and will later sort with the highest on top as follows:

    ```
    data.createOrReplaceTempView("houses");
    Dataset<Row>avgPrice = Spark.sql("select _c16 zip code,avg(_c2) avgPrice
    from houses group by _c16 order by avgPrice desc");
        avgPrice.show();
    ```

    And the result would be printed as:

    ```
    17/03/03 01:34:31 INFO CodeGenerator: Code generated in 36.
    +-------+------------------+
    |zipcode|          avgPrice|
    +-------+------------------+
    |  98039|         2160606.6|
    |  98004|1355927.0820189274|
    |  98040|1194230.0212765958|
    |  98112| 1095499.342007435|
    |  98102| 901258.2666666667|
    ```

For more elaborate data exploration and with graphs please refer to what we covered in *Chapter 2, First Steps in Data Analysis* and *Chapter 3, Data Visualization* and practice the code here.

Running and testing the linear regression model

For this example, we have used the dataframe machine learning API. It is an upcoming API from Apache Spark and it is built on the lines of the scikit-learn library. It is easy to use the machine learning algorithms on dataframe objects and they are massively scalable. Let's go through the code now.

We have the dataset now, so first we build the SparkSession and store it in a Spark object. For brevity I am not showing the full boilerplate code here:

```
SparkSession Spark = SparkSession.builder()...
```

Next load up the data from the dataset's CSV file (kc_house_data.CSV). After loading the dataset in the Spark dataset object we register it as a temporary view in Spark to fire queries on it:

```
Dataset<Row>fullData = Spark.read().CSV("data/kc_house_data.CSV");
   fullData.createOrReplaceTempView("houses");
```

Here you can do a fullData.printSchema() to see the schema (the column names, and so on) in the dataset. By default Apache Spark will give some columns names to the columns loaded in this CSV file and they will start with _c1, _c2, and so on.

Now filter out the columns that we need by firing a Spark SQL query:

```
Dataset<Row>trainingData = Spark.sql("select _c3 bedrooms,_c4
   bathrooms,_c5 sqft_living,_c6 sqft_lot,_c2 price from houses");
```

This is an important step. Note that our machine learning algorithm requires data in a particular form. Our machine learning algorithm from Apache Spark requires data to be in Dataset<Row> form only, but within the row object, the first value is the label or outcome of data (that is, if we are predicting the price so our outcome or label is price) and finally in the next value we pass a vector and this vector object is filled with the data of the features that we are feeding to our model.

Regression on Big Data

In the following code, we converted the dataframe to RDD first and later fired a map transformation on it. We invoke a Java `lambda` function and using that we are creating objects in the proper form as follows:

```
JavaRDD<Row>training = trainingData.javaRDD().map(s -> {
  return RowFactory.create(Double.parseDouble(s.getString(4).
  trim()),
    Vectors.dense(
Double.parseDouble(s.getString(0).trim()),

Double.parseDouble(s.getString(1).trim()),

Double.parseDouble(s.getString(2).trim()),

Double.parseDouble(s.getString(3).trim()))
        );
  });
```

> This model is a fully mathematical model and it needs to be fed with only numbers. Hence, we are converting all the feature values and the label outcomes to be numbers (double) first.
>
> If you have values in your datasets that are categorical or non numbers then before feeding them to your models you need to convert them to numbers first.

Since this is the new machine learning dataframe API, it has the dependency of the data to be present in `DataSet<Row>` format. Hence we convert our preceding training data RDD back to a `Dataset` object. First we defined the schema of the values that we have in our `JavaRDD`. Using this schema next we build the `Dataset` object by invoking `createDataFrame` on a Spark session:

```
StructType schema = new StructType(new StructField[]{
new StructField("label", DataTypes.DoubleType, false, Metadata.
empty()),
new StructField("features", new VectorUDT(), false, Metadata.empty())
});

Dataset<Row>trn = Spark.createDataFrame(training, schema);
  trn.show();
```

We have invoked a `trn.show()` here to show us the first few values in our dataset now. It will show up as a list of labels and features as follows:

```
+---------+--------------------+
|    label|            features|
+---------+--------------------+
| 221900.0|[3.0,1.0,1180.0,5...|
| 538000.0|[3.0,2.25,2570.0,...|
| 180000.0|[2.0,1.0,770.0,10...|
| 604000.0|[4.0,3.0,1960.0,5...|
| 510000.0|[3.0,2.0,1680.0,8...|
|1225000.0|[4.0,4.5,5420.0,1...|
```

We will split our dataset into two parts, one for training our model and the second part for testing our trained model as to how good it is. We keep 90 percent of the data for training and the remainder for testing:

```
Dataset<Row>[] splits = trn.randomSplit(newdouble[] {0.9, 0.1},
12345);
Dataset<Row>trainingMain = splits[0];
Dataset<Row>testMain = splits[1];
```

Now build our regression model and provide the parameters:

```
LinearRegression lr =
new LinearRegression().setMaxIter(50).setRegParam(0.3).
setElasticNetParam(0.5);
```

Train the model on the training data:

```
LinearRegressionModel lrModel = lr.fit(trainingMain);
```

Now test how good our training of the models is and how good the parameters we have chosen are. We will print the trained model coefficients and also print the root mean squared error. We will use the handy `LinearRegressionTrainingSummary` class that does these calculations for us and prints the root mean squared error:

```
System.out.println("Coefficients: "
   + lrModel.coefficients() + " Intercept: " + lrModel.intercept());
LinearRegressionTrainingSummary trainingSummary = lrModel.summary();
    System.out.println("RMSE: " + trainingSummary.
rootMeanSquareddError());
```

This would print the coefficients and root mean squared error as follows:

```
Coefficients: [-64534.41523975349,8313.685443179289,314.60167605061406,-0.37592410488893885] Intercept: 90080.21844645533
RMSE: 257059.12249611464
```

> As we discussed the root mean squared error before, we should try our model with different features and different parameters and see how our root mean squared error behaves. Our target is to reduce the root mean squared error, if it is going down then our model is getting better.

Finally, run our trained model on the test data split we had. After running the model using the `transform` function record the results of the model in the `rows` object, as shown in the following snippet, and collect and print the results:

```
Dataset<Row>results = lrModel.transform(testMain);
Dataset<Row>rows = results.select("features", "label", "prediction");
for (Row r: rows.collectAsList()) {
    System.out.println("(" + r.get(0) + ", " + r.get(1) + ") " + ", 
prediction=" + r.get(2));
}
```

This would print the results as follows:

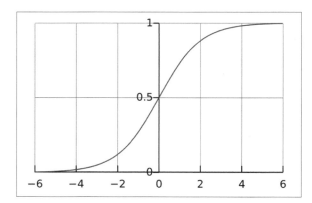

Look at the preceding printed results. A few values are predicted relatively well, for example, the second row from the top where the predicted results is quite near the actual value. Let's explore two values from this result:

Features and actual value	Predicted value	Description
[3.0,2.5,2390.0,6435.0], 432500.0	666740.1204811503	Values are widely off
[3.0,2.5,1650.0,2787.0], 433190.0	435306.25133833074	Values nearly match

As you can see in the preceding table, one value was close and the other was widely off. This is what machine learning is about; we need to get the best predicted results that are good for most cases. It takes years of practice on the part of data scientists to evaluate models that bring awesome predicted results. There are plenty of practical examples (for example, in the case of `Amazon.com` their website is full of predicted suggestive results that are very good) that are used in real life where the machine learning algorithms have been so finely optimized that they bring very good predictive results.

To fix our preceding results we need to try and test different feature combinations and do things such as cross-validation on the data and also tweak the parameters we pass to the model. Our aim in this book is to give you good introductory information on the predictive models and we would expect that the readers take it from there and practice with different models and different features and try to come up with the best results.

Next we will learn a very popular classification model, logistic regression.

Logistic regression

This is a popular classification algorithm where the dependent variable (outcome) is categorical. Even though it has the word regression in its name, it is a classification technique. Using this technique, we can train a model on some training data and the same model we can later use on new data to classify it into different categories. So, if you want to classify data into categories such as 1/0, Yes/No, True/False, Has Disease/No Disease, Sick/Not Sick and so on, logistic regression is a good classifier model to try in these cases. As per these examples, logistic regression is typically used for binary classification, but it can also be used for multiclass classification too.

The approach used by this algorithm is quite simple. We apply the data from the dataset onto a mathematical optimization function and this function will later make the data fall either in a 0 category or 1 category. Later on when we get a new piece of data we apply the same function to that new data and see where it falls, and based on that we predict its category.

Let's now try to dig deep into the concepts of logistic regression by asking a few questions.

Which mathematical functions does logistic regression use?

There are many mathematical functions that can be used with logistic regression. Basically, we will be feeding our parameters to this mathematical function and it will give an output of either 1 and 0, or 1 and -1. Based on this output we will be able to figure out which category is the output. We will be mainly discussing the sigmoid function.

A sigmoid is a mathematical function. It is defined by the following mathematical formula:

$$y = \frac{1}{1+e^{-x}}$$

Here:

- e: The natural logarithm base (also known as Euler's number)
- x: The x-value of the sigmoid's midpoint
- L: The curve's maximum value (you can see the curve in the following figure)
- k: The steepness of the curve (you can see the curve in the following figure)

This function is also called a sigmoidal curve. If it is plotted on the x and y axes it is an S-shaped curve shown as follows:

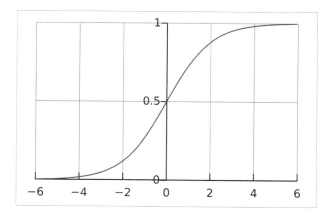

As seen in the preceding figure, it's an S-shaped graph and the graph starts from **-6** and goes to **+6** values. Its center lies at **0.5** (which is the value on the *y* axis). It's a very simple yet powerful function that states that if our value or outcome value is greater than 0.5 then we have the result as '1', and if it is less than 0.5 then the result is 0.

 Sigmoid is one of the functions we have covered here. There are other mathematical functions that are used in logistic regression, for example, the Heaviside Step function.

Thus logistic regression is basically an optimization technique that crunches your data to fall on either side of the boundary of a mathematical function. You can use this boundary condition, that is, greater or less than 0.5, as seen in the preceding graph, to figure binary classifications for labels such as sick/healthy, customer clicks the ad / customer avoids the ad, customer buys the product / customer ignores the product, and so on. In mathematical terms it is a probabilistic approach whereby the mathematical function figures out the probability of outcome and when it is beyond a certain value (as in our case it is 0.5) it is classified as one value or else the other value.

Logistic regression in use here is a binary classification technique where the output is of two kinds, but logistic regression can also be used as a multiple class classification technique. In a multiple class classification technique, the output can be of multiple kinds, for example, customer buys, ignores, or adds a product to their wish list. Thus, there are three categories here.

As we mentioned earlier, before feeding our features to the sigmoid function we multiply them with weights and then sum them up. But how do we calculate these weights? Calculation of proper weights or regression coefficients is important as based on that our logistic regression model performs. So naturally we get another question on this now. How do we calculate the regression coefficients?

We will briefly explain the two ways (though there are more ways) used to finding the regression coefficients or weights:

- **Gradient ascent or descent**: I won't go into the details of the math involved here much. For detailed reference on these approaches refer to the content on Wikipedia.

Gradient ascent is an iterative technique whereby we try to find the max point within a graph. To find the max point we will move in the direction of the gradient using some mathematical function. To move one step ahead every time along this path we would require a step function. We would try to minimize the loss or the cost function. As it is an iterative approach we can set up a maximum number of iterations beyond which this approach would stop. Similar to gradient ascent that finds a max function, a gradient descent finds a minimum function.

> After a required number of iterations are done and your model is ready run the predictive results and plot the results on a line that is superimposed on the scatter plots of the data. This will show you how well the data is partitioned using this approach.

- **Stochastic gradient descent**: The performance of the gradient ascent algorithm that was discussed previously is not very good because it re-evaluates the full data points on each run. To avoid this we use the stochastic gradient. This algorithm is very similar to the gradient descent one except that in this one we will update the weights or regression coefficients using only one instance at a time. For more information on this approach refer to Wikipedia.

Let's now try to find the uses of logistic regression.

Where is logistic regression used?

Logistic regression is a very useful technique. Here are some of its uses:

- Logistic regression is the base for more complex algorithms such as neural networks
- Logistic regression is used in medical sciences and social sciences, for example, whether a patient has a disease or not
- Logistic regression can be used in business too, for example, in e-commerce apps, it can be used to predict things such as whether the user will click on an ad or not, and so on

Predicting heart disease using logistic regression

Disease detection is an interesting problem and one that has been very active in the research field too. In most of the research papers on automatic disease detection machine learning is actively used to predict the occurrence of the disease based on various attributes. For this case study we will try to predict the occurrence of heart disease based on various parameters that the machine learning model is trained on. We will be using a heart disease dataset that is explained in the next section, for training our model.

> For the full steps of how we can run a machine learning model on big data, refer to *Chapter 4, Basics of Machine Learning,* where we talk about how data can be shipped into HDFS and how you can run the models using Spark and store the results in Parquet or other formats. Finally, the results can be consumed using products such as Impala to query the results in real time.

Dataset

The heart disease dataset is a very well-studied dataset by researchers in machine learning and it is freely available at the UCI machine learning dataset. Though there are 4 datasets in this, I have used the `Cleveland` dataset that has 14 main features; we are only using 5 of those. The dataset is in a CSV file and the features or attributes are as follows:

Feature	Description
Age	Age of the person
Sex	Male or female (1 = male; 0 = female)
cp	Chest pain type:
	value '1': typical angina
	value '2': atypical angina
	value '3': non-anginal pain
	value '4': asymptomatic
trestbpss	Resting blood pressure (in mm Hg on admission to the hospital)
chol	serum cholesterol in mg/dl
num	Diagnosis of heart disease (angiographic disease status):
	value 0: < 50% diameter narrowing (means 'No Disease')
	value 1: > 50% diameter narrowing (means 'Disease is Present')

Data cleaning and munging

The files contain data that has to be adapted into the format that the model requires. As it is a mathematical model it requires all numbers. In our dataset we have two main problems, as follows:

- **Missing data**: Some of the data points have null or no values and these would cause a null pointer exception to occur if we feed them to the model. In *Chapter 2, First Steps in Data Analysis,* we discussed a few strategies to deal with missing data, please refer to that chapter if you need to go through those details again. For now we will take a simplistic approach out of those approaches, we will replace the missing values with the mean value for that particular data column.

- **Categorical data**: The last parameter num has categorical string data in it. We can only feed numerical data to our model. For this we need to convert this string data to numbers. This categorical data signifies whether the user has a disease or not as such it can be depicted by binary values whereby 1 means has disease and 0 means no disease.

Data exploration

Let's do some initial data exploration on the data:

- **Number of items in the dataset**: It's a relatively small dataset and we will load the dataset and run a count on the number of rows:

    ```
    Dataset<Row>data = Spark.read().CSV("data/heart_disease_data.CSV");
    System.out.println("Number of Rows -->" + data.count());
    ```

    ```
    Number of Rows --> 303
    ```

- **Number of women and men in the dataset**: We will now find the total number of men and women in this dataset. For this, we will register the previous dataset as a temporary view and fire a Spark SQL group by query on it as follows:

    ```
    Dataset<Row>menWomenCnt = Spark.sql("select _c1 sex,count(*) from heartdiseasedata group by _c1");
    menWomenCnt.show();
    ```

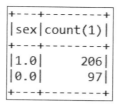

 In our dataset **1 = male** and **0 = female**.

- **Average age of women and men**: Let's find the average age of men and women in this dataset:

  ```
  Dataset<Row>menWomenAvgAge = Spark.sql("select _c1 sex,avg(_c0) from
  heartdiseasedata group by _c1");
      menWomenAvgAge.show();
  ```

  ```
  +---+------------------------+
  |sex|avg(CAST(_c0 AS DOUBLE))|
  +---+------------------------+
  |1.0|       53.83495145631068|
  |0.0|       55.72164948453608|
  +---+------------------------+
  ```

- **Minimum age of women and men with the disease**: Let's find the minimum age of women and men that have the disease:

  ```
  Dataset<Row>menWomenMinAge = Spark.sql("select _c1 sex,min(_c0) from
  heartdiseasedata group by _c1");
      menWomenMinAge.show();
  ```

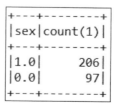

> We will leave it to the users to do more analysis on the data. As you can see, you can get valuable information just by exploration. For example, we saw that the minimum age for men with disease is 5 years less than women in this dataset; if we have more data than this dataset then even data exploration can give extremely useful results. Also you can plot graphs based on this data and the visualization of those graphs is quite insightful too.

So much for the data exploration piece; let's now dig into the logistic regression algorithm and see how it helps us predict the presence of heart disease given some data.

Running and testing the logistic regression model

The code for the model is very similar to what we showed for linear regression except that the type of model has changed in this case. We will start by building the `SparkSession` object, we call it `Spark` and later using that object, we will pull only the columns or features that we need to train our models with:

> Our model is a mathematical model and it feeds only on numbers. So make sure before feeding and training the model it has only numbers in the dataset.

```
SparkSession Spark = SparkSession.builder()...
```

Next, load up the data from the dataset's CSV file (`kc_house_data.CSV`). After loading the dataset in the Spark `Dataset` object, we register it as a temporary view in Spark to fire queries on it:

```
Dataset<Row>fullData = Spark.read().CSV("data/heart_disease_data.
CSV");
    fullData.createOrReplaceTempView("heartdiseasedata");
```

> Here you can do a `fullData.printSchema()` to see the schema (the column names, and so on) in the dataset. By default, Apache Spark will give some column names to the columns loaded in this CSV file and they will start with _c1, _c2, and so on.

Now filter out the columns that we need by firing a Spark SQL query:

```
Dataset<Row> selFeaturesdata = Spark.sql("select _c0 age,_c1 sex,_c2 
cp,_c3
sqft_lot, _c4 price,_c13 has_disease from heartdiseasedata");
```

After selecting the features that we need, we will convert all the features data into a number and put it into a vector object. Since our model requires a row object, we fill the row object with the outcome (from this training data) and the features into a vector as follows:

```
JavaRDD<Row>vectorsData = selFeaturesdata.javaRDD().map(s -> {
     return RowFactory.create((Double.parseDouble(s.getString(5).trim()) >
? 1.0 : 0.0, Vectors.dense(

Double.parseDouble(s.getString(0).trim()),
               Double.parseDouble(s.getString(1).trim()),
               Double.parseDouble(s.getString(2).trim()),
               Double.parseDouble(s.getString(3).trim()),
               Double.parseDouble(s.getString(4).trim()))

     );
});
```

Now we convert this JavaRDD back to a dataframe as our new dataframe machine learning API requires it to be in dataset format:

```
StructType schema = new StructType( new StructField[]{
new StructField("label", DataTypes.DoubleType, false, Metadata.empty()),
new StructField("features", new VectorUDT(), false, Metadata.empty())
   });

 Dataset<Row>trn = Spark.createDataFrame(vectorsData, schema);
   trn.show();
```

We have invoked a `trn.show()` here to show us the first few values in our dataset now. It will show up as a list of labels and features as follows:

```
+-----+--------------------+
|label|            features|
+-----+--------------------+
|  0.0|[63.0,1.0,1.0,145...|
|  1.0|[67.0,1.0,4.0,160...|
|  1.0|[67.0,1.0,4.0,120...|
|  0.0|[37.0,1.0,3.0,130...|
|  0.0|[41.0,0.0,2.0,130...|
```

We will split our dataset into two parts, one for training our model and the second part for testing our trained model as to how good it is. We keep 90 percent of the data for training and the remainder for testing:

```
Dataset<Row>[] splits = trn.randomSplit(newdouble[] {0.9, 0.1},
12345);
Dataset<Row>trainingData = splits[0];
Dataset<Row>testData = splits[1];
```

Now build our logistic regression model. We have checked that even with the default configuration this model performs well, so we have not done any extensive tweaking with the parameters here:

```
LogisticRegression lr = new LogisticRegression();
```

Now train the model on the training data:

```
LogisticRegressionModel lrModel = lr.fit(trainingData);
```

Now run our trained model on the test data split we had. After running the model using the transform function record the results of the model in the rows object, as shown in the following snippet, and collect and print the results:

```
Dataset<Row>results = lrModel.transform(testData);
Dataset<Row>rows = results.select("features", "label", "prediction");
for (Row r: rows.collectAsList()) {
    System.out.println("(" + r.get(0) + ", " + r.get(1) + ") " + ", 
prediction=" + r.get(2));
}
```

This would print the results as follows:

```
([35.0,1.0,2.0,122.0,192.0], 0.0) , prediction=0.0
([39.0,0.0,3.0,94.0,199.0], 0.0) , prediction=0.0
([47.0,1.0,3.0,130.0,253.0], 0.0) , prediction=0.0
([51.0,0.0,3.0,130.0,256.0], 0.0) , prediction=0.0
([51.0,1.0,3.0,125.0,245.0], 0.0) , prediction=0.0
([52.0,0.0,3.0,136.0,196.0], 0.0) , prediction=0.0
([52.0,1.0,3.0,138.0,223.0], 0.0) , prediction=0.0
([55.0,1.0,2.0,130.0,262.0], 0.0) , prediction=0.0
([56.0,0.0,2.0,140.0,294.0], 0.0) , prediction=0.0
([56.0,1.0,2.0,120.0,236.0], 0.0) , prediction=0.0
([60.0,0.0,3.0,120.0,178.0], 0.0) , prediction=0.0
([67.0,0.0,3.0,115.0,564.0], 0.0) , prediction=1.0
([67.0,0.0,3.0,152.0,277.0], 0.0) , prediction=0.0
([68.0,1.0,3.0,118.0,277.0], 0.0) , prediction=1.0
([42.0,1.0,4.0,136.0,315.0], 1.0) , prediction=1.0
([43.0,0.0,4.0,132.0,341.0], 1.0) , prediction=0.0
```

> False positive is a very dangerous predictive outcome in a disease dataset like this. A false positive states that the diseases are not present (as per the predictive results); however, in reality the disease is there. Thereby the false positive from a logistic regression model perspective is very bad for a predictive result. To counter this a little bit you can set a minimum probability beyond which only a certain result will be marked as positive or negative.

Now let's quickly check with a simple way as to how many of our predictions were good according to this model. For this we will take the total data rows in our test data and divide the number of wrong results with that and calculate the percentage of wrong results. The smaller the value the better our model is:

```
inttestDataLength = new Integer("" + rows.count());
intwrongResultsCnt = 0;
for (Row r: rows.collectAsList()) {
   if(r.getDouble(1) != r.getDouble(2)) wrongResultsCnt = wrongResultsCnt + 1;
}
doublepercentOfWrong = (wrongResultsCnt * 100)/testDataLength;
 System.out.println("Percent of wrong results -->" + percentOfWrong);
```

And the result is printed as follows:

```
Percent of wrong results --> 16.0
```

As you can see, our model gave 16% bad results. As such this model is not good for disease prediction. We need a much better accuracy if we need to make this model make some useful predictions. As for the readers, we would encourage the reader to try different models on the same dataset and observe the error rate for practice.

Summary

In this chapter, we studied two very popular machine learning algorithms, namely linear regression and logistic regression. We saw how linear regression can be used to predict continuous values such as sales counts, estimating the price of products, and so on. We also ran a sample case study using the linear regression approach to predict the prices of houses. We later learned about logistic regression and ran a sample using a popular heart disease dataset used for studying machine learning.

In the next chapter, we will learn two more supervised learning algorithms that are used heavily in classification. The first algorithm that we will study is Naive Bayes and then we will learn about the support vector machine algorithm.

6
Naive Bayes and Sentiment Analysis

A few years back one of my friends and I built a forum where developers could post useful tips regarding the technology they were using. I wished I knew about the Naive Bayes machine learning algorithm then. It could have helped me to filter objectionable content that was posted on that forum. In the previous chapter, we saw two algorithms that can be used to predict continuous values or to classify between discrete sets of values. Both the approaches predicted a definite value (whether it was continuous or discrete), but they did not give us a probability of occurrences of our best guesses. Naive Bayes gives us the predicted results with a probability attached to it, so in a set of results for same category we can pick the one with the highest probability.

In this chapter, we will cover:

- General concepts about probability and conditional probability. This section will be basic and users who already know this can skip this section.
- We will cover the bayes theorem upon which the Naive Bayes algorithm is based.
- We will look into the concepts of Naive Bayes and see some real-life use cases.
- After this we will use a simple example to understand the concepts of a Naive Bayes algorithm.
- Finally, we will run a real-world sample case study on a Twitter dataset for sentimental analysis. For this we will be using a standard machine learning algorithm, and the big data toolsets such as Apache Spark, HDFS, parquet, Spark ML, and Spark SQL.

> While doing sentiment analysis (we will explain this in detail within this chapter) we will cover a lot of features on text analysis on top of big data.

- Before we get into the details of the Naive Bayes algorithm we must understand the concepts of conditional probability.

Conditional probability

Conditional probability in simple terms is the probability of occurrence of an event given that another event has already occurred. It is given by the following formula:

$P(B|A) = P(A \text{ and } B)/P(A)$

Here in this formula the values stand for:

Probability value	Description
$P(B|A)$	This is the probability of occurrence of event B given that event A has already occurred.
$P(A \text{ and } B)$	The probability that both event A and B occur.
$P(A)$	This is the probability of occurrence of an event A.

Now let's try to understand this using an example. Suppose we have a set of seven figures as follows:

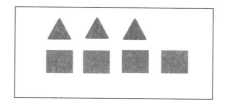

As seen in the preceding figure, we have three triangles and four rectangles. So if we randomly pull one figure from this set the probability that it belongs to either of the figures will be:

$P(triangle)$ = Number of Triangles / Total number of figures = 3 / 7

$P(rectangle)$ = Number of rectangles / Total number of figures = 4 / 7

Now suppose we break the figure into two individual sets as follows:

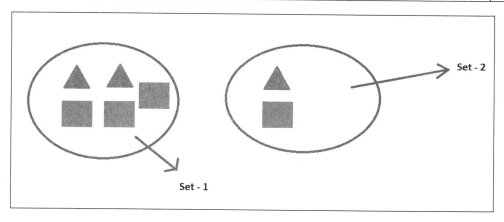

As seen in the preceding figure, we have two sets each with their own individual figures. Now suppose we pull a figure from **Set-1** then what is the probability that it is a triangle? There are two events involved here, the first event is the event of pulling a figure from the **Set-1** and the second event is that the set chosen is **Set-1**. We know that the set chosen is **Set-1** as described by the problem hence the second event has already occurred. Thus we can write the formula as follows:

P(triangle and set-1) = Number of triangles in set-1 / Total figures = 2 / 7

P(set-1) = Number of figures in set-1 / total number of figures = 5 / 7

Therefore,

P(triangle | set-1)= P(triangle and set-1) / P(set-1) = (2/7) / (5/7) = 2/5 = 0.4

Thus, our conditional probability of pulling a triangle from **Set-1** is 0.4. We have seen what conditional probability is, and now it's time to learn the bayes theorem.

Bayes theorem

The Bayes theorem is based on the concept of learning from experience, that is, using a sequence of steps to come to a prediction. It is the calculation of probability based on prior knowledge of occurrences that might have led to the event. Bayes theorem is given by the following formula:

$$P(A|B) = \frac{P(B|A)P(A)}{P(B)}$$

Where:

Probability Value	Description
P(A | B)	Conditional probability of event A given that event B has occurred.
P(B | A)	Conditional probability of event B given that event A has occurred.
P(A)	Individual probability of event A without regard to event B.
P(B)	Individual probability of event B without regard to event A.

Let's understand this using the same example as we used previously. Suppose we picked one green triangle randomly from a set then what is the probability that it came from **Set-1**?

Before we run the bayes theorem formula we will first calculate the individual probabilities:

- Probability of randomly picking a set from one of the two sets, Set-1 and Set-2

    Since there are two sets only this probability is

    $$\frac{1}{2} = 0.5$$

- Probability of picking a triangle from any set

    $$P(Triangle) = \frac{Number\ of\ Triangles}{Total\ Figures\ Count} = \frac{3}{7} = 0.43$$

- Probability of picking a triangle given that it came from Set-1

    $$P(Triangle\ |\ Set\_1) = \frac{P(Triangle\ and\ Set_1)}{P(Set_1)} = \frac{2}{5} = 0.4$$

- Now we will go to our original bayes formula and find the probability we are looking for, that is

    $$P(Set\_1\ |\ Triangle) = \frac{P(Triangle\ |\ Set\_1)\ P(Set\_1)}{P(Triangle)} = \frac{0.4 * 0.5}{0.43} = 0.47$$

    Thus, the probability that the triangle came from *Set_1* is 0.47 or 47%.

- To find that the triangle came from *Set_2* is simple as it would be the remaining probability and we can delete it from 1. Thus

    $$P(Set\_2\ |\ Triangle) = 1 - P(Set_1\ |\ Triangle) = 1 - 0.47 = 0.53$$

So the probability that the triangle came from *Set_2* is *0.53 or 53%*.

- As the probability of picking the triangle from *Set_2* is higher than the probability of picking it from Set-1, thus we can predict that the triangle came from *Set_2*.

Before we knew which set the triangle belonged to, the probability of pulling the triangle was 40% or 0.4. But once we knew what bucket the figures belonged to, the probability increased to 53% when it is pulled from *Set_1*. This makes sense too right, once we know which set the bucket belongs to it becomes a little easier to figure out whether it's a triangle or not, and hence the increase in probability. And this is what bayes theorem is all about, where a new probability is figured out as a consequence of a set of sequences (which in our case was depicting which set the figure belonged to).

Also, if you closely look at the preceding results you would realize that we just solved a **classification** problem. Thus, we gave a few attributes (figure type, that is, triangle) and asked the user to classify between two discrete values, that is, *Set_1* and *Set_2*.

For a full blown classification problem we will have many more features as compared to the single feature (the figure type, that is, triangle or rectangle) shown previously.

We will next look at the Naive Bayes algorithm that is based on this probability principle of bayes theorem.

Naive Bayes algorithm

Have you ever wondered how your Gmail application automatically figures out that a certain message that you have received is spam and automatically puts it in the spam folder? Behind the email spam detector, a powerful machine learning algorithm is running, that automatically detects whether a particular email that you have received is spam or useful. This useful algorithm that runs behind the scenes and saves you wasted hours on deleting or checking these spam emails is Naive Bayes. As the name suggests, the algorithm is based on the bayes theorem. The algorithm is simple yet powerful, from the perspective of classification the algorithm figures out the probability of occurrence of each discrete class and it picks the value with the highest probability.

You might have wondered why the algorithm carries the word Naive in its name. It's because the algorithm makes some Naive assumptions that the features that are present in a dataset are independent of each other. Suppose you have an email that contains the word casino and gambling, this algorithm will make the assumption that both these words are completely independent and occurrence of one will not affect the occurrence of the other one in any way. Even in this example of casino and gambling this assumption looks wrong since we all know that people go to casinos mainly for gambling. Even though Naive Bayes makes such an assumption of features being independent of each other the overall performance of the algorithm is still pretty good and hence it is used in many real-life applications such as spam filters. In the next section, we will see some of the real-life applications of this algorithm.

Naive Bayes is used in a lot of practical real-life applications as follows:

- This algorithm is the base for many spam filters and spam classifiers. As such it is used in many popular email clients, forum software, comments filtering software, and so on for filtering spam content.
- It is used in sentimental analysis of text to classify the emotion of a particular piece of text (for example, a review of a product) whether it is a positive emotion or a negative one.
- For document categorization, for example, to classify an article into categories such as politics, sport, technology, and so on.
- This algorithm is fast to train and test; hence it is used in real-time prediction scenarios to make fast predictions on events based data that is generated in real time.
- It is used in many recommendation systems to give useful suggestions of content to the users.
- We have seen some of the real-life use cases of Naive Bayes; we will now learn some of its advantages and disadvantages.

Advantages of Naive Bayes

Even though Naive Bayes looks like a simple algorithm, it can give tough competition to many of the similar popular machine learning algorithms. Some of its major advantages are:

- It is simple and fast to train. It can be trained on smaller sets of data too. Hence if you are looking for something very fast to try and train Naive Bayes is a good choice to go for first.

- Since it can be trained fast it is also useful in real-time prediction systems as well. For example, in terms of big data if you have a real-time event transfer system such as Kafka that is hooked to transfer data around, you can predict on this data in real time using the Naive Bayes algorithm.
- If the assumption of independent features holds good on the dataset that you are working on, and then it can give good competition to other machine learning algorithms such as logistic regression in terms of performance.
- It can make both binary and multi class classification.
- It can be trained to work in parallel (as you can see its implementation in Apache Spark also works in parallel), and do to this it can easily scale on massive datasets especially in the case of big data datasets.

We have seen the advantages of this algorithm; let's now look at some of its drawbacks.

Disadvantages of Naive Bayes

We have seen that Naive Bayes makes some strong assumptions on the dataset features, as such, it has some drawbacks too. We will explore some of those drawbacks now:

- Naive Bayes assumes that the features in the dataset are completely independent of each other. This goes well in some datasets where the features are relatively independent, but in datasets where the features are tightly coupled or related, it can give bad performance in terms of predictions. Suppose you like sugary drinks and salty drinks, the Naive Bayes can predict well whether you will like a drink or not when given a sugary drink or salty drink. But suppose you don't like drinks that contain both sugar and salt, then in this case Naive base would predict badly as it considers both the features independent and cannot relate both of the features together.
- If there is a response variable in the test data, but correspondingly it does not have value in the training data then it would assign it a probability of zero. In this case it won't be able to make a prediction. To solve this, a value is added to the zero probability and this is called a **smoothing factor** (it is called a Laplace Estimation).

We have now gone through some of the basic concepts of the Naive Bayes algorithm, and we have seen that this algorithm is extensively used in text analysis. So we will try to understand its usage by looking at a real-life example of sentimental analysis, which is a form of text analysis. Before diving directly into the algorithm, we will study some of the basic concepts regarding sentimental analysis.

Sentimental analysis

As we showed in the previous examples, Naive Bayes has extensive usage in text analysis.

One of the forms of text analysis is **sentimental analysis**. As the name suggests this technique is used to figure out the sentiment or emotion associated with the underlying text. So if you have a piece of text and you want to understand what kind of emotion it conveys, for example, anger, love, hate, positive, negative, and so on you can use the technique sentimental analysis. Sentimental analysis is used in various places, for example:

- To analyze the reviews of a product whether they are positive or negative
- This can be especially useful to predict how successful your new product is by analyzing user feedback
- To analyze the reviews of a movie to check if it's a hit or a flop
- Detecting the use of bad language (such as heated language, negative remarks, and so on) in forums, emails, and social media
- To analyze the content of tweets or information on other social media to check if a political party campaign was successful or not

Thus, sentimental analysis is a useful technique, but before we see the code for our sample sentimental analysis example, let's understand some of the concepts needed to solve this problem.

> For working on a sentimental analysis problem we will be using some techniques from natural language processing and we will be explaining some of those concepts.

Concepts for sentimental analysis

Before we dive into the fully-fledged problem of analyzing the sentiment behind text, we must understand some concepts from the **NLP** (**Natural Language Processing**) perspective.

We will explain these concepts now.

Tokenization

From the perspective of machine learning one of the most important tasks is feature extraction and feature selection. When the data is plain text then we need some way to extract the information out of it. We use a technique called **tokenization** where the text content is pulled and tokens or words are extracted from it. The token can be a single word or a group of words too. There are various ways to extract the tokens, as follows:

- **By using regular expressions**: Regular expressions can be applied to textual content to extract words or tokens from it.
- **By using a pre-trained model**: Apache Spark ships with a pre-trained model (machine learning model) that is trained to pull tokens from a text. You can apply this model to a piece of text and it will return the predicted results as a set of tokens.

To understand a tokenizer using an example, let's see a simple sentence as follows:

Sentence: "The movie was awesome with nice songs"

Once you extract tokens from it you will get an array of strings as follows:

Tokens: ['The', 'movie', 'was', 'awesome', 'with', 'nice', 'songs']

> The type of tokens you extract depends on the type of tokens you are interested in. Here we extracted single tokens, but tokens can also be a group of words, for example, 'very nice', 'not good', 'too bad', and so on.

Stop words removal

Not all the words present in the text are important. Some words are common words used in the English language that are important for the purpose of maintaining the grammar correctly, but from conveying the information perspective or emotion perspective they might not be important at all, for example, common words such as **is**, **was**, **were**, **the**, and **so**. To remove these words there are again some common techniques that you can use from natural language processing, such as:

- Store stop words in a file or dictionary and compare your extracted tokens with the words in this dictionary or file. If they match simply ignore them.
- Use a pre-trained machine learning model that has been taught to remove stop words. Apache Spark ships with one such model in the Spark feature package.

Let's try to understand stop words removal using an example:

Sentence: "The movie was awesome"

From the sentence we can see that common words with no special meaning to convey are **the** and **was**. So after applying the stop words removal program to this data you will get:

After stop words removal: ['movie', 'awesome', 'nice', 'songs']

 In the preceding sentence, the stop words **the**, **was**, and **with** are removed.

Stemming

Stemming is the process of reducing a word to its base or root form. For example, look at the set of words shown here:

car, cars, car's, cars'

From our perspective of sentimental analysis, we are only interested in the main words or the main word that it refers to. The reason for this is that the underlying meaning of the word in any case is the same. So whether we pick car's or cars we are referring to a car only. Hence the stem or root word for the previous set of words will be:

car, cars, car's, cars' => car (stem or root word)

For English words you can again use a pre-trained model and apply it to a set of data for figuring out the stem word. Of course there are more complex and better ways (for example, you can retrain the model with more data), or you have to totally use a different model or technique if you are dealing with languages other than English. Diving into stemming in detail is beyond the scope of this book and we would encourage readers to check out some documentation on natural language processing from Wikipedia and the `Stanford nlp` website.

 To keep the sentimental analysis example in this book simple we will not be doing stemming of our tokens, but we will urge the readers to try the same to get better predictive results.

N-grams

Sometimes a single word conveys the meaning of context, other times a group of words can convey a better meaning. For example, 'happy' is a word in itself that conveys happiness, but 'not happy' changes the picture completely and 'not happy' is the exact opposite of 'happy'. If we are extracting only single words then in the example shown before, that is 'not happy', then 'not' and 'happy' would be two separate words and the entire sentence might be selected as positive by the classifier However, if the classifier picks the bigrams (that is, two words in one token) in this case then it would be trained with 'not happy' and it would classify similar sentences with 'not happy' in it as 'negative'. Therefore, for training our models we can either use a unigram or a bigram where we have two words per token or, as the name suggests, an n-gram where we have 'n' words per token, it all depends upon which token set trains our model well and improves its predictive results accuracy.
To see examples of n-grams refer to the following table:

Sentence	The movie was awesome with nice songs
Uni-gram	['The', 'movie', 'was', 'awesome', 'with', 'nice', 'songs']
Bi-grams	['The movie', 'was awesome', 'with nice', 'songs']
Tri-grams	['The movie was', 'awesome with nice', 'songs']

For the purpose of this case study we will be only looking at unigrams to keep our example simple.

By now we know how to extract words from text and remove the unwanted words, but how do we measure the importance of words or the sentiment that originates from them? For this there are a few popular approaches and we will now discuss two such approaches.

Term presence and Term Frequency

Term presence just means that if the term is present we mark the value as 1 or else 0. Later we build a matrix out of it where the rows represent the words and columns represent each sentence. This matrix is later used to do text analysis by feeding its content to a classifier.

Term Frequency, as the name suggests, just depicts the count or occurrences of the word or tokens within the document. Let's refer to the example in the following table where we find term frequency:

Sentence	The movie was awesome with nice songs and nice dialogues.
Tokens (Unigrams only for now)	['The', 'movie', 'was', 'awesome', 'with', 'nice', 'songs', 'and', 'dialogues']
Term Frequency	['The = 1', 'movie = 1', 'was = 1', 'awesome = 1', 'with = 1', 'nice = 2', 'songs = 1', 'dialogues = 1']

As seen in the preceding table, the word 'nice' is repeated twice in the preceding sentence and hence it will get more weight in determining the opinion shown by the sentence.

Bland term frequency is not a precise approach for the following reasons:

- There could be some redundant irrelevant words, for example, **the, it**, and **they** that might have a big frequency or count and they might impact the training of the model
- There could be some important rare words that could convey the sentiment regarding the document yet their frequency might be low and hence they might not be inclusive for greater impact on the training of the model

Due to this reason, a better approach of TF-IDF is chosen as shown in the next sections.

TF-IDF

TF-IDF stands for **Term Frequency** and **Inverse Document Frequency** and in simple terms it means the importance of a term to a document. It works using two simple steps as follows:

- It counts the number of terms in the document, so the higher the number of terms the greater the importance of this term to the document.
- Counting just the frequency of words in a document is not a very precise way to find the importance of the words. The simple reason for this is there could be too many stop words and their count is high so their importance might get elevated above the importance of real good words. To fix this, TF-IDF checks for the availability of these stop words in other documents as well. If the words appear in other documents as well in large numbers that means these words could be grammatical words such as **they, for, is**, and so on, and TF-IDF decreases the importance or weight of such stop words.

Let's try to understand TF-IDF using the following figure:

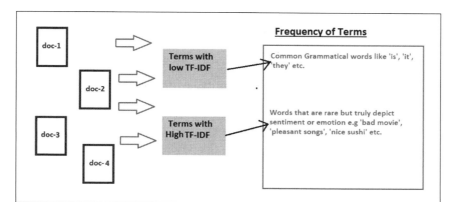

As seen in the preceding figure, **doc-1**, **doc-2**, and so on are the documents from which we extract the tokens or words and then from those words we calculate the TF-IDFs. Words that are stop words or regular words such as **for**, **is**, and so on, have low TF-IDFs, while words that are rare such as 'awesome movie' have higher TF-IDFs.

TF-IDF is the product of Term Frequency and Inverse document frequency. Both of them are explained here:

- **Term Frequency**: This is nothing but the count of the occurrences of the words in the document. There are other ways of measuring this, but the simplistic approach is to just count the occurrences of the tokens. The simple formula for its calculation is:

    Term Frequency = Frequency count of the tokens

- **Inverse Document Frequency**: This is the measure of how much information the word provides. It scales up the weight of the words that are rare and scales down the weight of highly occurring words. The formula for inverse document frequency is:

    $$Inverse\ Document\ Frequency = \log \frac{Total\ Number\ of\ Documents}{Number\ of\ Documents\ containing\ the\ Term}$$

- **TF-IDF**: TF-IDF is a simple multiplication of the Term Frequency and the Inverse Document Frequency. Hence:

    *TF-IDF = Term Frequency * Inverse Document Frequency*

This simple technique is very popular and it is used in a lot of places for text analysis. Next let's look into another simple approach called bag of words that is used in text analytics too.

Bag of words

As the name suggests, bag of words uses a simple approach whereby we first extract the words or tokens from the text and then push them in a bag (imaginary set) and the main point about this is that the words are stored in the bag without any particular order. Thus the mere presence of a word in the bag is of main importance and the order of the occurrence of the word in the sentence as well as its grammatical context carries no value. Since the bag of words gives no importance to the order of words you can use the TF-IDFs of all the words in the bag and put them in a vector and later train a classifier (Naive Bayes or any other model) with it. Once trained, the model can now be fed with vectors of new data to predict on its sentiment.

We have seen the steps that we will be using for sentimental analysis, let's now start digging into the code. We will look at our dataset first.

Dataset

Our dataset contains a single file with lots of movie reviews and the corresponding sentiment, that is, whether the review is positive or negative. The file contains data that is tab separated. Some of the first few lines from the dataset are shown here:

```
1    The Da Vinci Code book is just awesome.
1    i liked the Da Vinci Code a lot.
1    I loved the Da Vinci Code, but now I want something better and different!..
1    i liked the Da Vinci Code a lot.
1    I liked the Da Vinci Code but it ultimately didn't seem to hold it's own.
```

As seen in the preceding screenshot, the dataset is mostly text data and text data is generally huge in size if you are trying to pull text from sources such as social media, `log` files, and so on. As such for text analysis a big data stack is best. Storing the text data in HDFS is a good option as HDFS is highly scalable. Our dataset file size is not huge, but to mimic a big data environment let's now put the file in HDFS using the following command:

```
hdfs -dfs put <Filename><HDFS_DIR_NAME>
```

As we can see, we use the `hdfs` command to put the file from the operating system file system to the `hdfs` directory.

We will now do some general data exploration on the file.

Data exploration of text data

To explore our model we will first load our dataset from HDFS. To do this we will first create the `SparkSession` (using Spark configuration) and then load the text file using the `sparkContext` as follows. We will not show the boiler plate code though the full code can be seen in the GitHub package for this book:

```
SparkConf c = ...

SparkSession spark = ...

JavaRDD<String> data =
spark.sparkContext().textFile("hdfs://data/sa/training.txt",
1).toJavaRDD();
```

As we can see, we load a `JavaRDD` object with text data that is loaded from a `textFile` in `hdfs`.

> The data exploration piece is not tied to any specific package from Spark. So we should feel free to use both the RDD API or Spark dataset API for our exploration.

Next, we fire a map function on this `JavaRDD` and the `map` function is then applied to each row of data. Each row is a sentence within the dataset along with a label of sentiment. From this row of data the sentiment and the sentence are extracted (they are tab separated) and stored in a Java POJO object called `TweetVO`. The `JavaRDD` object is now a distributed list of these POJOs:

```
JavaRDD<TweetVO>tweetsRdd = data.map(strRow -> {
    String[] rowArr = strRow.split("\t");
    String realTweet = rowArr[1];

        TweetVOtvo = new TweetVO();
            tvo.setTweet(realTweet);
            tvo.setLabel(Double.parseDouble(rowArr[0]));
            returntvo;
    });
```

Let's now find the number of rows in our dataset:

```
System.out.println(" Numbers of Rows --> " + tweetsRdd.count());
```

This would print the result as follows:

```
Number of Rows --> 7086
```

Next we create a dataframe out of our RDD using the following code:

```
Dataset<Row>dataDS = spark.createDataFrame(tweetsRdd.rdd(), TweetVO.class);
```

After creating the dataframe let's see the first few lines of our data:

```
dataDS.show(5);
```

This would print the first five lines of data as follows:

```
17/07/12 09:00:30 INFO CodeGenerator:
+-----+--------------------+
|label|               tweet|
+-----+--------------------+
|  1.0|The Da Vinci Code...|
|  1.0|this was the firs...|
|  1.0|i liked the Da Vi...|
|  1.0|i liked the Da Vi...|
|  1.0|I liked the Da Vi...|
+-----+--------------------+
only showing top 5 rows

17/07/12 09:00:30 INFO SparkContext:
```

We will now count the number of positive labels versus the number of negative labels or sentiments in the dataset we have. For this first we register our dataset as a temporary view and then fire an SQL query on it using the group by function to count the labels:

```
dataDS.createOrReplaceTempView("tweets");
Dataset<Row> saCountDS = spark.sql("select label sentiment, count(*) from tweets group by label");
saCountDS.show();
```

This would print the output as follows:

```
17/07/12 09:07:03 INFO CodeGenerator:
+---------+--------+
|sentiment|count(1)|
+---------+--------+
|      0.0|    3091|
|      1.0|    3995|
+---------+--------+

17/07/12 09:07:03 INFO SparkContext:
```

As seen here, the number of positive reviews (depicted by 1) are more than the number of negative review (depicted by 0).To view the count of labels in a better way we will plot this on a bar chart. For this we will reuse the SQL query that we just depicted previously and fill the results of this query into a chart object (specific to JFreeChart). We will create an instance of DefaultCategoryDataset used by the bar charts and later we will collect the data from the saCountDS dataset created previously. We will iterate over this data and from each data row we will extract the label and the corresponding count value of the sentiment and fill it into the defaultCategoryDataset object:

```
finalDefaultCategoryDataset dataset = new DefaultCategoryDataset(
);
List<Row> results = saCountDS.collectAsList();
    for (Row row : results) {
            String key = "" + row.getDouble(0);
            if(null == key) key = "(Empty Values)";
            else if("1.0".equals(key))   key = "Positive";
            else key = "Negative";
            dataset.addValue(row.getLong(1) , category , key );
    }

            return dataset;
```

For maintaining the brevity of the code we are not showing the code for the full chart here. For the full code refer to the code in our GitHub repository. Also you can refer to the previous *Chapter 5, Regression on Big Data* that we covered on charts.

This would then create the chart as follows:

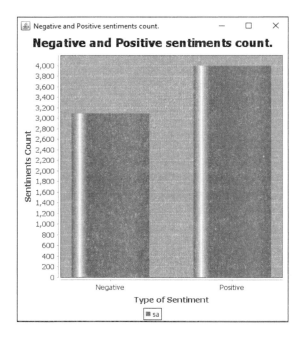

As seen in the preceding bar chart, the number of positive sentiments is more than the number of negative sentiments.

As we have mostly text data, let's now see how many words there are in our dataset and we will count these number of words. Also we will try to sort the top words.

To create a word count program:

1. We first create our dataset of data in the same way as we did initially in our data exploration section:

   ```
   Dataset<Row>tweetsDs = spark.createDataFrame(tweetsRdd.rdd(),
   TweetVO.class);
   ```

   After loading the dataset we create a `Tokenizer` instance. This class is provided in the Spark ML package and it is a pre-trained model to extract tokens from textual content. In our case, it would extract words from each row of sentences. As seen in the following code, we provide the column where the `tokenizer` would read the data from and the column where it would output its results:

   ```
   Tokenizer tokenizer = new Tokenizer().setInputCol("tweet").
   setOutputCol("words");
   ```

2. Next we run this `tokenizer` and store its results in a new `Dataset` object:

   `Dataset<Row>dataDs = tokenizer.transform(tweetsDs);`

3. Next we extract the column that has the results of our `tokenizer`.

4. After extracting the column we convert this dataset to a `JavaRDD` object and invoke a `flatMap` function on it to flatten the list of words and store each word in a string per row of this new `JavaRDD`:

   `JavaRDD<Object> words = dataDs.select("words").javaRDD().flatMap(s ->s.getList(0).iterator() );`

5. Next we invoke a `mapToPair` function on this words `JavaRDD` so as to create a pair of each word with the default value of 1 with the word itself as the key:

   `JavaPairRDD<String, Integer>wpairs = words.mapToPair(w -> new Tuple2(w.toString(), 1) );`

6. Finally, we invoke a `reduceByKey` to sum up the count of words:

   `JavaPairRDD<String, Integer>wcounts = wpairs.reduceByKey((x,y) -> x + y);`

7. To analyze the results, we convert this RDD back to dataset so that we can fire SQL queries on it. For this on the `wcounts` pair function we invoke a `map` function and fill the results in `WordVO` POJO:

   ```
   JavaRDD<WordVO>wordsRdd = wcounts.map(x -> {
               WordVOvo = new WordVO();
                   vo.setWord(x._1);
                   vo.setCount(x._2);
                   returnvo;
   });
   ```

8. Next we create a dataframe out of this `wordsRdd` and register it as a temporary view with the name as `words`. We are not showing this code for brevity. On this words view we now fire a query to collect the top words used, as shown in the following snippet:

   ```
   Dataset<Row>topWords = spark.sql("select word,count from words order by count desc");
                   topWords.show();
   ```

This would print the result as follows:

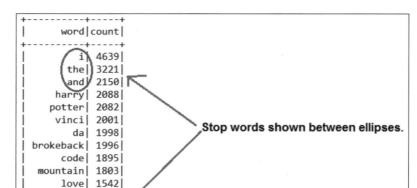

It's no wonder that our top words are stop words such as **i**, **the**, **and**, and so on. These words don't add much for sentimental analysis and as such should be removed.

> We leave it for the users to fire further queries on our dataset and try out other things such as changing the case of the words, removing all the special characters, trying out different token types like bigrams, and so on.

So much for the data exploration piece let's now dive into the actual code for sentimental analysis.

Sentimental analysis on this dataset

In this program, we will do the following steps for sentimental analysis, as shown in the following figure:

Chapter 6

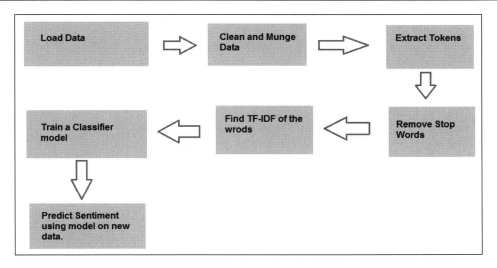

As seen from the steps in the preceding diagram, our simple approach is:

- Load the dataset from HDFS in Spark
- Clean and filter the data from the dataset
- On each row of data use a tokenizer to parse the tokens
- Remove the stop words from the tokens
- Find the TF-IDF of the words and feed it to a classifier
- Finally test the model on some new data for sentiment prediction

 Our approach is very simplistic and there is no rocket science approach. Our aim is to just get the concepts laid out in front of the users. In the real world, sentimental analysis is much more complicated and is an active research area.

Finally, we will jump into the code now. Let's build the boiler plate code first for the `SparkSession`:

```
SparkConf c= ...
SparkSession spark= ...
```

 For brevity we are not showing all the boiler plate code here for building the `SparkSession`.

[175]

Next, we load the dataset from `hdfs` and store it in an RDD of strings:

```
JavaRDD<String> data = spark.sparkContext().textFile("hdfs://data/sa/
training.txt", 1).toJavaRDD();
```

Fire a `map` function on this RDD and extract data per row and store it in a POJO (`TweetVo` class in our case):

```
JavaRDD<TweetVO>tweetsRdd = data.map(strRow -> {
    String[] rowArr = strRow.split("\t");
    String rawTweet = rowArr[1];
    String realTweet = rawTweet.replaceAll(",", "").replaceAll("\"",
"").replaceAll("\\*", "").replaceAll("\\.", "").trim();
        TweetVOtvo = new TweetVO();
            tvo.setTweet(realTweet);
            tvo.setLabel(Double.parseDouble(rowArr[0]));
            returntvo;
});
```

This code is similar to what we did in data exploration. The main point is we have the data cleaning code also here where we remove the special characters from the data. The raw data is part of the `rawTweet` object and we clean it to build the `realTweet` object.

Remember we are using the new Spark ML package, which runs the machine learning algorithms on the Spark dataframe. But we loaded our data into an `rdd` class, hence we convert it into a dataframe now. To do so we invoke the spark `createDataFrame` method and pass it the `rdd` as well as the POJO class:

```
Dataset<Row>tweetsDs = spark.createDataFrame(tweetsRdd.rdd(), TweetVO.
class);
```

Now we have the dataset ready to use for our training and testing. We break this dataset into the individual training and testing datasets:

```
Dataset<Row>[] tweetsDsArr = tweetsDs.randomSplit(new double[]
{0.8,0.2});
Dataset<Row> training = tweetsDsArr[0];
Dataset<Row> testing = tweetsDsArr[1];
```

Each row of our dataset is essentially an English sentence. Now create a `tokenizer` and provide the input column to read the data from and an output column where the `tokenizer` would store the new list of words after removing stop words:

```
Tokenizer tokenizer = new Tokenizer().setInputCol("tweet").
setOutputCol("words");
```

If you individually execute this `tokenizer` and collect its output you will be able to see its output as follows:

```
#############---- START ---- #############
Opinion --> 0.0
Sentence --> 00 we rode bikes to hollywood and rented brokeback mountain which was also stupid
Tokens --> [00, we, rode, bikes, to, hollywood, and, rented, brokeback, mountain, which, was, also, stupid]
#############---- END ---- #############

#############---- START ---- #############
Opinion --> 0.0
Sentence --> 1-BROKEBACK MOUNTAIN IS A STUPID MOVIE
Tokens --> [1-brokeback, mountain, is, a, stupid, movie]          Tokens
#############---- END ---- #############

#############---- START ---- #############
Opinion --> 0.0
Sentence --> 10 Things I Hate About You + A Knight's Tale  Brokeback Mountain =
Tokens --> [10, things, i, hate, about, you, +, a, knight's, tale, , brokeback, mountain, =]
#############---- END ---- #############
```

As you can see, the reviews that are fed to the movie are broken into individual words. The arrow shows the words that are tokenized.

Next, initialize a stop words remove object using this feature from the Spark ML library. Again, provide the input and output columns:

```
StopWordsRemoverstopWrdRem =
new StopWordsRemover().setInputCol("words").
setOutputCol("updatedWords");
```

If we run this stop words remover individually and collect its output it will print the results as follows:

```
#############---- START ---- #############
Opinion --> 0.0
Sentence --> 00 we rode bikes to hollywood and rented brokeback mountain which was also stupid
Tokens --> [00, we, rode, bikes, to, hollywood, and, rented, brokeback, mountain, which, was, also, stupid]
After removing Stop Words --> [00, rode, bikes, hollywood, rented, brokeback, mountain, also, stupid]
#############---- END ---- #############

#############---- START ---- #############
Opinion --> 0.0
Sentence --> 1-BROKEBACK MOUNTAIN IS A STUPID MOVIE
Tokens --> [1-brokeback, mountain, is, a, stupid, movie]
After removing Stop Words --> [1-brokeback, mountain, stupid, movie]
#############---- END ---- #############

#############---- START ---- #############
Opinion --> 0.0
Sentence --> 10 Things I Hate About You + A Knight's Tale  Brokeback Mountain =
Tokens --> [10, things, i, hate, about, you, +, a, knight's, tale, , brokeback, mountain, =]
After removing Stop Words --> [10, things, hate, +, knight's, tale, , brokeback, mountain, =]
#############---- END ---- #############
```

As you can see in the preceding screenshot, the ellipses depict the stop words that have been removed by the stop words remover feature of the Spark ML API.

As is the case with most machine learning algorithms in the Apache Spark package, they work on an input set of features that are in the form of a feature **vector**. Due to this we need to put our extracted words into a vector form. For this we will use a `HashingTF` (though you can use CountVectorizer too) class provided by Apache Spark. HashingTF is a Term Frequency generator and it creates a vector filled with term or token frequencies. It is an optimized algorithm as it uses a hashing technique for this; please refer to official Spark documentation for more details. As in other cases the pattern is similar we provide the input and output column except in this case we provide the number of features too:

```
intnumFeatures = 10000;
HashingTFhashingTF = new HashingTF()
                    .setInputCol("updatedWords")
                    .setOutputCol("rawFeatures")
                    .setNumFeatures(numFeatures);
```

> HashingTF is similar to a hast table that stores the words as a key and their counts except that it maintains a distributed structure. The previous number of features depicts the bucket size of this hash table. The more the buckets, the less the collisions and more words it can store.

We can use these term frequencies to train our simplistic model, but that won't be good simply because the frequency of some useless words could be very high. To better gauge at the importance of real good words we calculate their TF-IDF using this vector of term frequencies created by HashingTF:

```
IDFidf = new IDF().setInputCol("rawFeatures").
setOutputCol("features");
```

After calculating our Inverse Document Frequencies we need to train our model that it can build its probability levels based on these frequencies to figure out the sentiment of the text. For this we train our Naive Bayes model. We provide the input column where the model can read the TF-IDF frequencies vector and the output column where it can store the predicted sentiment:

```
NaiveBayesnb = newNaiveBayes().setFeaturesCol("features").setPredictio
nCol("predictions");
```

Finally, use the `Pipeline` API to hook all this together:

```
Pipeline p = newPipeline();
```

Provide all the steps of the workflow `tokenizer`, stop words removal, and so on to this `Pipeline` object:

```
p.setStages(new PipelineStage[]{ tokenizer, stopWrdRem, hashingTF,
idf,nb});
```

Provide the training set of data to run through the workflow using the `Pipeline` object and train the model:

```
PipelineModelpm = p.fit(training);
```

Finally, run the trained model on the testing dataset and store the predictions in a dataset object:

```
Dataset<Row>updTweetsDS = pm.transform(testing);
        updTweetsDS.show();
```

This will print the first few lines of the predictions as follows:

```
+--------------------+--------------------+--------------------+-----------+
|            features|       rawPrediction|         probability|predictions|
+--------------------+--------------------+--------------------+-----------+
|(10000,[1702,2007...|[-460.73982498224...|[1.0,2.8831832778...|        0.0|
|(10000,[493,2007,...|[-664.19237642734...|[0.99999999970660...|        0.0|
|(10000,[1871,4420...|[-301.86338122585...|[0.99999999984274...|        0.0|
|(10000,[2007,2069...|[-520.19457361779...|[2.80856472431555...|        1.0|
|(10000,[1083,1402...|[-154.19765901965...|[1.0,1.7444214330...|        0.0|
|(10000,[1083,1402...|[-154.19765901965...|[1.0,1.7444214330...|        0.0|
|(10000,[1083,1402...|[-154.19765901965...|[1.0,1.7444214330...|        0.0|
```

To fit the content to page we are not showing all the columns in the prediction results. The last columns show the predictions whether positive or negative, that is, either 1 or 0.

As Naive Bayes works on the principle of probability, there is an additional column for probability added onto the dataset that contains the conditional probability of each feature (word).

So much for the predictive results. Let's now check how good our trained model is. For this we will use the `MulticlassClassificationEvaluator` class provided by the Spark framework and create an instance of this class. We will also provide the actual label and the predicted label to this class along with the metric we are interested in (in our case it is `accuracy`):

```
MulticlassClassificationEvaluator evaluator =
newMulticlassClassificationEvaluator()
            .setLabelCol("label")
            .setPredictionCol("predictions")
            .setMetricName("accuracy");
```

Next we calculate the accuracy value by invoking an `evaluate` method of the `evaluator` on our dataset:

```
double accuracy = evaluator.evaluate(updTweetsDS);
System.out.println("Accuracy = " + accuracy);
System.out.println("Test Error = " + (1.0 - accuracy));
```

This will print the output as follows:

```
Accuracy = 0.9753086419753086
Test Error = 0.024691358024691357
```

The accuracy comes out to 0.97 or 97%, which is not bad at all for our simplistic Naive Bayes classifier.

> The approach we showed here is very basic. We would urge the users to try a different set of features, clean up the data further, and retest the models to improve the accuracy.

We have now seen one popular algorithm called Naive Bayes, let's now briefly learn another popular machine learning algorithm in the next section.

SVM or Support Vector Machine

This is another popular algorithm that is used in many real life applications like text categorization, image classification, sentiment analysis and handwritten digit recognition. Support vector machine algorithm can be used both for classification as well as for regression. Spark has the implementation for linear SVM which is a binary classifier. If the datapoints are plotted on a chart the SVM algorithm creates a hyperplane between the datapoints. The algorithm finds the closest points with different labels within the dataset and it plots the hyperplane between those points. The location of the hyperplane is such that it is at maximum distance from these closest points, this way the hyperplane would nicely bifurcate the data. To figure out this maximum distance for the location of the hyperplane the SVM algorithm uses a `kernel` function (mathematical function).

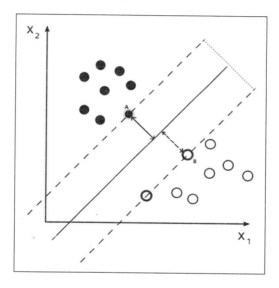

As you can see in the image we have two different type of datapoints one clustered on the **X2** axis side and the other clustered on the **X1** axis side. There is a unique plane that separates the two closest points marked as **A** and **B** in the image. The hyperplane which is actually a straight line in this two-dimensional image is the solid line and the distance is shown by the perpendicular lines from the point **A** and point **B**. The dashed lines which pass through point **A** and point **B** are another hyperplanes which are not good as they do not segregate the data well. To figure out the location of this hyperplane a mathematical kernel function is used. To learn more about Support Vector Machine please refer to its documentation on Wikipedia.

From Spark MLlIB 2.2 onwards there is a complete linear support vector machine algorithm that is bundled inside the Spark machine learning library itself. We will now use that linear support vector machine algorithm from Spark and do the same sentiment analysis piece that we did earlier using Naive Bayes. Almost the entire code that we discussed in the previous section using Naive Bayes algorithm would remain the same and just the portion where we actually use the Naive Bayes model would change.

As shown in the code we will build an instance of a `LinearSVC` model which is a binary support vector machine classifier bundled inside Spark library. To this instance we supply the necessary parameter of the name of the column where the algorithm can read the vectorized features and the name of the column where the algorithm can put the predicted results. In our case these columns are `features` and `predictions`

```
LinearSVC linearSVM = new
LinearSVC().setFeaturesCol("features").setPredictionCol("predictio
ns");
```

After creating the instance of the model the remaining code is exactly the same as what we did for Naive Bayes. We will put this model into the flow for `Pipeline` API of Spark and remove the old Naive Bayes model that we used earlier here as shown.

```
Pipeline p = new Pipeline();
    p.setStages(new PipelineStage[]{ tokenizer, stopWrdRem,
    hashingTF, idf, linearSVM});
```

Next we will just fit this pipeline model on the training data and run it on the test data and finally verify the accuracy of our model. Since this code is exactly similar to what we showed in the previous section please refer to the previous section for this code.

The output of the dataset that stores the predictions from the SVM model is as shown:

```
17/07/13 00:12:44 INFO Coder
+-----+--------------------+--------------------+--------------------+-----------+
|label|               tweet|            features|       rawPrediction|predictions|
+-----+--------------------+--------------------+--------------------+-----------+
|  0.0|10 Things I Hate ...|(10000,[2007,2069...|[1.73011104770074...|        0.0|
|  0.0|A couple of very ...|(10000,[2007,3189...|[1.24141630788482...|        0.0|
|  0.0|A futile mission ...|(10000,[2345,3326...|[-0.2558815088181...|        1.0|
|  0.0|A mother in Georg...|(10000,[161,294,4...|[2.93919879710469...|        0.0|
|  0.0|AND BROKEBACK MOU...|(10000,[493,2007,...|[1.51404977982603...|        0.0|
|  0.0|After school I we...|(10000,[1871,4420...|[1.90088379349260...|        0.0|
|  0.0|After the festivi...|(10000,[437,2007,...|[1.99976959365757...|        0.0|
```

The output of the accuracy of the SVM classifier is as shown:

```
17/07/13 00:12:45 INFO DAGScheduler: Job 87
Accuracy = 0.9846796657381616
Test Error = 0.015320334426183843
17/07/13 00:12:45 INFO SparkContext: Invoki
```

As shown in the result of the SVM classifier the performance of the classifier has improved further and it is now 0.98 or 98 %. Support vector machine is a popular model in many real world applications as it performs well. For more information on this model please refer to Wikipedia.

Summary

This chapter covered a lot of ground on two important topics. Firstly, we covered a popular probabilistic algorithm, Naive Bayes, and explained its concepts and showed how it uses bayes rule and conditional probability to make predictions about new data using a pre-trained model. We also mentioned why Naive Bayes is called Naive as it makes a Naive assumption that all its features are completely independent of each other, thereby occurrence of one feature does not impact the other in any way. Despite this it forms well as we saw in our sample application. In our sample application we learnt a technique called sentimental analysis for figuring out the opinion whether positive or negative from a piece of text.

In the next chapter, we will study another popular machine learning algorithm called decision tree. We will show how it is very similar to a flowchart and we will explain it using a sample loan approval application.

7
Decision Trees

Decision trees are one of the simplest (and most popular) of machine learning algorithms, yet they are extremely powerful and used extensively. If you have used a flowchart before, then understanding a decision tree won't be at all difficult for you. A decision tree is a flowchart except in this case, the machine learning algorithm builds this flowchart, for you. Based on the input data, the decision tree algorithm automatically internally creates a knowledge base of a set of rules based on which it can predict an outcome when given a new set of data. In this chapter, we will cover the following topics:

- Concepts of a decision tree machine learning classifier, including what a decision tree is, how it is built, and how it can be improved
- The uses of the decision tree
- A sample case study using decision trees for classification

Let's try to understand the basics of decision trees now.

What is a decision tree?

A decision tree is a machine learning algorithm that belongs to the family of supervised learning algorithms. As such, they rely on training data to train them. From the features on the training data and the target variable, they can learn and build their knowledge base, based on which they can later take decisions on new data. Even though decision trees are mostly used in classification problems, they can be used very well in regression problems also. That is, they can be used to classify between discrete values (such as 'has disease' or 'no disease') or figure out continuous values (such as the price of a commodity based on some rules).

As mentioned earlier, there are two types of decision trees:

- **Decision trees for classification**: These are the decision tree algorithms that are used in classification of categorical values, for example, figuring out whether a new customer could be a potential loan defaulter or not.
- **Decision trees for regression**: These are the decision tree algorithms that are used in the predicting continuous values, for example, what size loan (in amount) can be given to a particular new customer based on certain criteria or attributes.

Let's try to understand a decision tree using the perspective of a flowchart. In a flowchart, we go from one flow to another based on rules, for example, if an event has occurred, we choose one direction or the other. Similar to that, in a decision tree, we have a bunch of rules (that are created by our machine learning classifier) and that direct our direction of decision flow. Understanding a decision tree becomes easy when we look at some examples. We will now try to understand a decision tree using a simple example.

Suppose there are two people applying for a job position at Java in big data analytics. Let's call these people Candidate A and Candidate B. Now suppose the candidates have the following skillsets:

- **Candidate-A**: Java, big data, web applications
- **Candidate-B**: Java, big data, analytics on big data

Let's build a simple decision tree to evaluate which candidate is suitable for the Java position, considering that only the candidate who has the desired skillset will be selected.

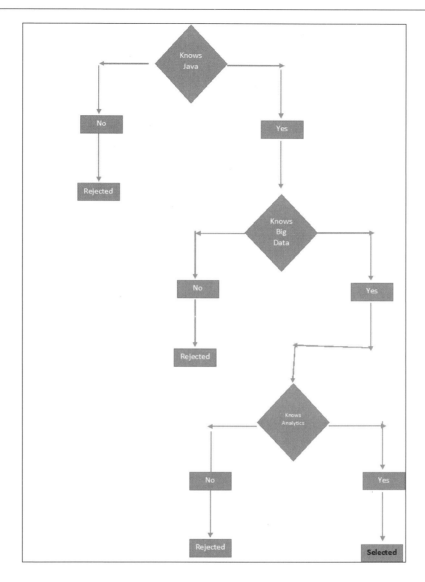

As you can see in the preceding diagram, the final decision—or candidate selected—is **Candidate-B**, as he is the one whose skillset matches the desired skillset.

This is as simple as a decision tree can be. It is just like a flowchart or tree structure built on a set of rules. Upon following the rules from the top of the tree to the leaf node, we can figure out the outcome and classify the end result. From the perspective of machine learning, we build a model that is trained on a set of features and labels (labels are the final outcome for these features). The model learns from these features and builds its knowledge base for the set of rules needed for the tree. Based on these rules, the model can then predict decisions on a new set of data.

Decision trees have been effectively used in a lot of industries. Here are some real-life examples of decision trees:

- In risk analysis in financial systems
- In software systems to estimate the development efforts on software modules and for other purposes
- In astronomy to discover galaxy counts, filtering noise from telescope images, and so on
- In medical research for diagnosis, cardiology, and so on
- For text classification, building personal learning assistants, to classify sleep patterns, and so on

As you saw in the preceding diagram, visualizing a decision tree is easy and helps us easily figure out what an outcome will be. A decision tree built by a human being is one thing, as we can visualize nicely, but a decision tree built by a computer program is complex and requires a few techniques utilized from pure mathematics. Next, we will study how a computer program builds a decision tree.

Building a decision tree

Up until recently, decision trees were one of the most used machine learning algorithms. They have been used extensively with ensembled algorithms (we will explain ensembling in the next chapter). Building a decision tree involves programs that can read data from a dataset and then split the data into sections based on a rule. This rule—or split rule—is figured out using mathematical techniques. To decide on which feature is best suited to the split, the algorithm will split on every feature and will pick the feature that makes elements within the two individual split sets most similar to each other. Let's try to understand this using an example:

Suppose you have a dataset of fruits with the following attributes:

Color	Diameter	Fruit (target variable)
Red	4	Apple
Orange	4	Orange
Red	1	Cherry
Red	5	Apple
Orange	3	Orange

Now, our task is to build a decision tree with this dataset of attributes (color and diameter) and based on these attributes the decision algorithm should be able to figure out the type of fruit when a new data point is given to it with a specific color and diameter. As you can see, there are two features or attributes in this dataset **Color** and **Diameter**. We will evaluate both cases when we split the dataset based on these features as shown next:

- **Splitting by color = 'red'**: Though there are two color types in this dataset (**red** and **orange**), for brevity we are showing only the first **red** color type. We will split the dataset based on this color and figure out how homogeneous the split sets of data turn out to be.

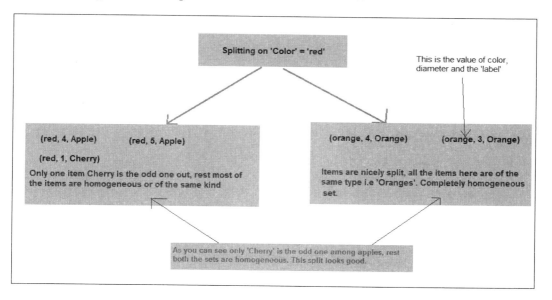

Decision Trees

As you can see in the preceding diagram, the split is quite good. Only **cherry** is the item on the left-hand side that is out of place otherwise the split criteria has nicely split the values and both the split sets are homogeneous or almost homogeneous (that is, of the same type)

- **Splitting by diameter > 3**: Here I have used the value 3 because 3 is the average diameter of the items. The split will be as shown in the following diagram:

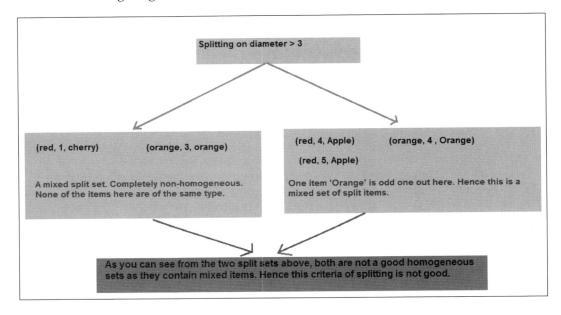

As you can see in the preceding diagram, this is not a very good split as it results in mixed sets. We want splits that are as homogeneous as possible.

For us human beings, visualizing a decision tree is easy and we can easily plot it on a diagram using our human intelligence. But what about a computer program—how does a computer program figure out the set of rules it should bifurcate decisions on? We will now see how a computer program builds a decision tree based on the data it is trained with.

Choosing the best features for splitting the datasets

The accuracy of a tree depends upon how good a split it does. The best feature chosen to split upon will directly impact the prediction results. There are mathematical ways by which we can decide upon the split criteria and the mathematical functions to use. The general approach that is employed to find the best split feature is:

- First split the dataset by each feature one at a time
- Next, record how homogeneous each split set is by figuring out the homogeneousness using a mathematical function such as entropy or Gini impurity
- The feature that splits the datasets into most homogeneous split sets is the best feature for the split and is chosen

The question that remains is, "How do we measure the homogeneousness of the split sets?" To do so, we use two mathematical approaches called Gini Impurity and Entropy. Now we will explain them in detail:

1. **Gini Impurity**: This is a measure of the probability of an item being correctly labeled when it is randomly picked from a split set. If you have a split set with k discrete classes of samples, then Gini Impurity is given by the following formula:

$$Gini\ Impurity = 1 - \sum_{i=1}^{k} P_i^2$$

Here, P_i is the probability of picking that class of sample from the split set.

Now, suppose we split on some criteria and our split set contains the following data as shown in the following table:

Color	Diameter	Fruit (target variable)
Red	4	Apple
Orange	4	Orange
Orange	3	Orange
Red	5	Apple
Orange	3	Orange

Thus, our split set has 3 Oranges and 2 Apples, which will be calculated as follows:

$P(oranges) = 3/5$

$P(apples) = 2/5$

$$Gini\ Impurity = 1 - \left[\left(\frac{3}{5}\right)^2 + \left(\frac{2}{5}\right)^2\right] = 0.48$$

We showed one such split earlier. What if the decision tree algorithm makes multiple splits based on a feature?

In that case, we calculate the Gini Impurity of all the split sets. We later find the net Gini Impurity of all the split sets using the following formula:

$$Net\ Gini\ of\ Split\ Sets = \sum_{i}^{n} \frac{elements\ in\ set}{Total\ Elements} * Gini\ of\ Set$$

After finding the Gini of all the split sets, we find the GiniGain. This property tells how much impurity is reduced by making the split. This is a simpler difference between the Gini of the parent set and the Net *Gini of the Split Sets*. Thus, we can say that:

GiniGain = Gini of the Parent Set - Gini of the split sets

The decision tree uses that split, which results in producing the maximum GiniGain.

1. Entropy: In simple terms, Entropy is the measure of mixedness or impurity in a collection of examples. If there is a collection of examples with all the items of the same type, then it is completely pure. Since, if you randomly pull any item from it, it will be of the one type only. A pure collection like this has an entropy of 0. Similarly, if there is a set with all the items of different types, then it has one hundred percent impurity and the entropy will be 1. The aim of the decision tree split is to generate a new set of split values where this amount of Entropy is reduced. This concept of reduction of entropy based on a split criterion (or rule) is called as information gain.

2. If you have a set with *xss* different values in it, then the formula for Entropy calculation is given by:

$$\text{Entropy(X)} = -\sum_{i=1}^{n} P(x_i) \log_b P(x_i)$$

As seen in the preceding formula, if *i* is one such item in the *set(x)* then we find its probability within the set and multiply it by the logarithmic of its probability (log to the base 2). We do this for all the different discrete classes within the impure set and sum up all the calculated values for each different class in the set.

Let's understand this with a simple example.

Suppose you have a set of different fruits such as:

Set(fruits) = {apples, apples, apples, orange, orange, orange, orange, orange}

On this set, we find the probability of pulling each fruit.

So, it will be calculated as follows:

Probability(apple) = 3/8

Probability(orange) = 5/8

So, the Entropy for the Set *(fruits)* will be:

$$Entropy(fruits) = -\left[\left(\frac{3}{8}\right)\log_2\frac{3}{8} + \left(\frac{5}{8}\right)\log_2\frac{5}{8}\right] = 0.95$$

Using this Entropy calculation, our decision tree would calculate the information gain on each split. So specifically what the decision tree would do is shown in the following diagram:

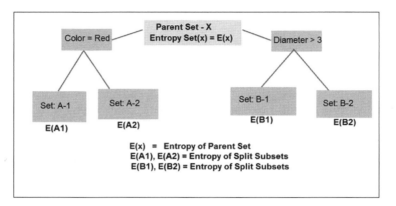

As you can see in the preceding diagram, the decision tree would perform the following steps:

1. First, the decision tree algorithm calculates the *Entropy* of a *Parent Set*. In our case, as shown in the image, it is $E(x)$.
2. Next, it splits the set based on a feature, for example, here we split on two features separately one feature is **Color** and the other is **Diameter**.
3. After splitting, it finds the entropy of each split set. In the diagram, the split sets Entropy is $E(A1)$, $E(A2)$, $E(B1)$, and $E(B2)$.
4. Finally, it finds the net Entropy of the split sets. This is calculated as follows:

$$\text{Net Entropy of Split Sets} = \sum_{i}^{n} \frac{elements\ in\ set}{Total\ Elements} * Entropy\ of\ Set$$

Here, *i* represents one Split Set and n the total number of Split Sets.

5. Finally, the information gain is calculated as the difference between the parent Entropy and the new net split sets Entropy. The more the value the better the split is. So the information gain in the preceding example (in the image) is calculated for both split criteria (that is, color and diameter both) and the one with the higher value is selected.

Even though we have discussed only two approaches, we encourage readers to check more information on other approaches for splitting such as Chi-square and Gini index.

Building a perfect tree is computationally very expensive and hard at the same time. To counter this, our task is to come up with a good enough tree that suits our need.

 A good enough tree is suited to more conditions, as in most applications you would use techniques such as ensembling to use a group of trees instead of a single tree algorithm. Hence, a combination of all these trees to figure out the predictive results will ultimately yield an excellent solution as we will see in the next chapter.

Advantages of using decision trees

Decision trees are very popular machine learning models and we will now study some of the advantages of using them:

- As you have seen, they are very simple to build and use. They are essentially a set of if...else statements that lead to a conclusive result.
- The input can be of any type whether numeric or strings for checking the decision type.
- From the perspective of big data and distributed computing, it's easier to build a decision tree model that can be distributed on a cluster of machines. Thus, it can run parallely and can be very fast.

Apart from these advantages, decision trees also suffer from some problems. Let's look at some of these problems now.

Disadvantages of using decision trees

All the machine learning algorithms come with a few pros and cons. Decision trees are no exception to this. Let's look at some of the disadvantages of using decision trees:

- Decision trees suffer from the problem of overfitting. Overfitting is a generic problem with many machine learning models. Due to overfitting, the models get well acquainted to the data that they are trained with and they perform extremely well on the training data, but the same models perform poorly on any new data that was not part of the training set.

- If there are too many decision rules, then the model can soon become quite complex. Since we are dealing with big data, this problem is more common.

 You might have observed by now that as the number of decision rules increases more number of splits are required and hence the amount of computations required also increases. This will slow down the whole process of decision tree computations.

Dataset

A loan approval dataset is a sample dataset that is freely available on the web globally please. This is just a sample dataset, which contains rows of data with various attributes. The outcome or response of each row shows whether the loan application was approved or rejected. The attributes in each row of the dataset are shown in the following table:

Attribute name	Description
Loan ID	This states the Loan ID, which is a unique variable
Gender	This states the gender—male or female
Married	This is the marital status—married or unarried
Dependents	This states the number of dependents for the person
Education	This states the educational qualification
Self employed	This states whether the person is self-employed
Applicant income	This states the income of the loan applicant
Coapplicant income	This states the income of the co-applicant of the loan
Loan amount	This states the amount of the loan
Loan amount term	This states the duration of the loan
Credit history	This states the credit history of the person applying for loan (good or bad)
Property area	This states the area of property in case this is a housing loan
Loan status	This states whether the application was approved or rejected

Data exploration

Before we dive into running our models for training and testing the dataset, let's explore the dataset first for understanding the data. For this, we will first create the Spark session instance and we will load our dataset from the dataset file. For brevity, we will not show the boilerplate code for `SparkSession` creation.

```
SparkSession spark = …
Dataset<Row> rowDS = spark.read().csv("data/loan/loan_train.csv");
```

Let's see the first few rows of our dataset by running the `show()` method:

```
rowDS.show()
```

This will give us the following output:

```
+--------+----+---+---+------------+---+----+----+----+---+---+---------+---+
|     _c0| _c1|_c2|_c3|         _c4|_c5| _c6| _c7| _c8|_c9|_c10|     _c11|_c12|
+--------+----+---+---+------------+---+----+----+----+---+---+---------+---+
|LP001002|Male| No|  0|    Graduate| No|5849|   0|null|360|  1|    Urban|  Y|
|LP001003|Male|Yes|  1|    Graduate| No|4583|1508| 128|360|  1|    Rural|  N|
|LP001005|Male|Yes|  0|    Graduate|Yes|3000|   0|  66|360|  1|    Urban|  Y|
|LP001006|Male|Yes|  0|Not Graduate| No|2583|2358| 120|360|  1|    Urban|  Y|
|LP001008|Male| No|  0|    Graduate| No|6000|   0| 141|360|  1|    Urban|  Y|
|LP001011|Male|Yes|  2|    Graduate|Yes|5417|4196| 267|360|  1|    Urban|  Y|
|LP001013|Male|Yes|  0|Not Graduate| No|2333|1516|  95|360|  1|    Urban|  Y|
|LP001014|Male|Yes| 3+|    Graduate| No|3036|2504| 158|360|  0|Semiurban|  N|
|LP001018|Male|Yes|  2|    Graduate| No|4006|1526| 168|360|  1|    Urban|  Y|
```

As you can see, Spark automatically names the columns as `_c0`, `_c1`, and so on. You can map this dataset to a schema and probably to some good column names. For us, we will be using the same column names for our queries.

Next we will find the total rows in this dataset:

```
rowDS.createOrReplaceTempView("loans");
```

This would print the output as follows:

```
Number of rows à 768
```

As you can see, it's a small dataset, but for learning purposes this is sufficient for now.

Decision Trees

We will now find the number of males and females in the dataset:

```
Dataset<Row> maleFemaleDS = spark.sql("select _c1 gender,count(*) cnt
from loans group by _c1");
        maleFemaleDS.show();
```

This would print the result as follows:

As you can see, there are thirteen null values in the gender column and this has to be fixed. We will replace this with the median value in the column (of course, we will round the value to the nearest number). Similar to this, you can group and find the count of other columns too, this is a handy way for finding null value counts or garbage values apart from the real good values. But Apache Spark has a great API and it provides a very handy method for checking the details of your columns and here is that method:

```
rowDS.describe("_c0","_c1","_c2","_c3","_c4","_c5","_c6","_c7","_c8","_c9","_c10","_c11","_c12").show();
```

And this would print the summary of all the columns including their count (where the values were found) and their mean value.

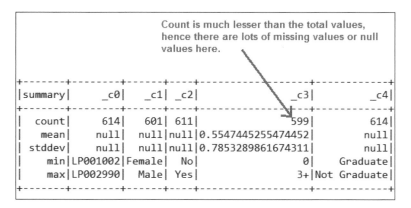

For maintaining brevity, we are not showing all the columns here. But we need to fix all the columns with null values or discard these columns completely and not use them in training our models. We will cover data cleaning in our next section.

[198]

Chapter 7

Let's now plot the loan amount on a box chart to get some information regarding the loan amount values (such as loan amount median values) to see if any outlier points are present. Refer to *Chapter 3, Data Visualization* where we saw how to make a box chart.

As shown in the following code, we just need to properly pull the data from the dataset using Apache Spark and extract the loan amount data (that is, _c6 column) from it.

```
Dataset<Row> rowDS = spark.read().csv("data/loan_train.csv");
  rowDS.createOrReplaceTempView("loans");
Dataset<Row> loanAmtDS = spark.sql("select _c6 from loans");
```

Next, pull the data from its dataset and populate it into a list object:

```
List<Row> loanAmtsList = loanAmtDS.collectAsList();
List list1 = new ArrayList<Double>();
  for (Row row : loanAmtsList) {
    if( null != row.getString(0) &&
    !"".equals(row.getString(0)))
  list1.add(Double.parseDouble(row.getString(0)) );
  }
```

Finally, use this list object to populate a default `DefaultBoxAndWhiskerCategoryDataset` and return this dataset. The chart will be printed as shown next:

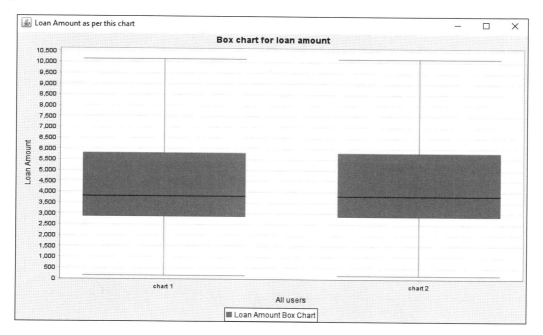

Decision Trees

When it comes to loans, one of the most important parameters is credit history. Let's now analyze how many loans were approved for people when they had good credit history and bad credit history. For this, we draw a bar chart for the approved loans and plot the credit history on the *x* axis as shown.

 For the code for the bar chart generation, refer to our code samples in the GitHub repository. In order to maintain brevity, we are not showing the code here.

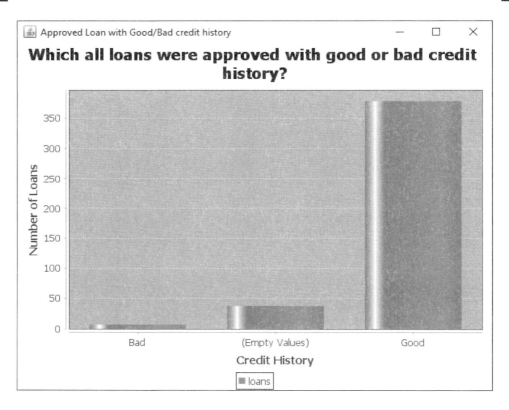

There are two important things you can make out from the preceding graph. The first is that most of the loans with good credit were approved and very few were approved for those people with bad credit history. Also, there are around fifty (or slightly less) data points that do not have any value on the credit history column. So we would need special data cleaning operations on these empty column values. Also, similar to the preceding chart for approved loans, we can make another chart for rejected loans in a similar way. We can also make bar charts or histograms on the other values such as loan amounts and applicant income in a similar way to understand our data better.

Apart from giving some good details about the data such as which parameters impact loan approval like credit history, what is the average applicant income is, the number of males versus females in our dataset, we also saw that our dataset is quite unclean. We need to fix our data by adding the missing values before we can apply any model to it for training. In our next section, we will show how to clean and mung this data.

Cleaning and munging the data

As our dataset is pretty unclean and it has lots of missing values, we need to fix these values before we can train our models. As we mentioned in *Chapter 2*, *First Steps in Data Analysis* there are various ways to add the missing values such as replacing them with the mean value of the column, ignoring the missing points altogether or using the k-NN (k-nearest neighbour) algorithm and picking the most similar data point value next to the missing one, we will be using a simple approach here, by taking a mean of the column values and replacing the missing values with the mean value.

We will fix the missing values of the credit history first. As credit history has two discrete values, we are taking the approach of finding the percentage of values a positive credit history and with the percentage of values with a negative credit history. Considering that we have the `SparkSession` object ready, let's fire a Spark SQL query to figure out the count of credit history values with good and bad credit history:

```
Dataset<Row> rowAvgDS = spark.sql("select _c10,count(*) from loans where _c10 is
not null group by _c10 ");
```

In the preceding code, the `_c10` column is for credit history and this would print the following result:

```
+----+--------+
|_c10|count(1)|
+----+--------+
|   0|      89|
|   1|     475|
+----+--------+
```

As such 0 represents bad credit history and 1 represents good credit history. If we find the percentage of good versus bad credit history in the non-null data rows, we will see that close to 85% of rows have positive credit history. Thus, we will put positive credit history or 1 in the missing ones. Of course we could have tried more fancy ways such as using a machine learning model to predict the best missing value in this case.

Decision Trees

We will now figure out the value that we want to put for the missing loan amounts. Here we will take a simple approach as we will just replace the value with the average or median value for the loan amounts. To find the average or mean value of the loan amount, you can run a simple Spark SQL query as shown next.

Here, we are running a Spark SQL query on the loans dataset and invoking on the `avg` function from Spark SQL to find the average value of loan amount based on the loan amount fields that are not null in the dataset.

```
Dataset<Row> avgLoanAmtRow = spark.sql("select avg(_c8) avgLoanAmount
from loans where _c8 is not null");
    avgLoanAmtRow.show();
```

This would print the following result:

If we round off the preceding value to the nearest number, the value comes to `146`. So, wherever we see the value missing for the average loan amount, we replace it with this value. We will show this piece in the code in the next section. Also, similar to the loan amount, we can replace the missing values of applicant income with the average value of applicant income.

> We only showed fixing the missing features loan amount, applicant income, and credit history, but there are many other features in this dataset. Even for training our models we are only going to use these features. We will leave it for the readers as an exercise to work on the remaining features for data cleaning and training their models if they want to use them to predict loan approvals.

Next, we will look at the actual training and testing of our decision tree model.

Training and testing the model

For this particular algorithm, we will use the new pipeline API from Apache Spark. The advantage of using the Apache Spark pipeline API is, it nicely encapsulates our workflow, or the entire steps for training and testing our model. This is useful in the maintenance of large-scale machine learning implementations.

The general steps of training and testing the model, however, will be similar to what we did earlier. That is:

- Loading the dataset
- Cleaning and munging the data in the dataset
- Extracting features from the dataset
- Training the various steps in the pipeline workflow, including feature extraction, model training, and model testing
- Running the model using the pipeline object and collecting the results
- Finally, evaluate how good our model is and improve its accuracy by trying different sets of features

Let's dive into the code now. Start the program by writing the boilerplate code to build the `SparkSession`. We will refer to this object as `spark`:

```
SparkSession spark = ...
```

Next, using this `spark` object, we load data from the CSV file containing our data:

```
Dataset<Row> rowDS = spark.read().csv("data/loan/loan_train.csv");
```

After loading the data into our dataset, we register it as a temporary view so that we can fire SQL queries on it:

```
rowDS.createOrReplaceTempView("loans");
```

Before we do actual training of your models, we must do the data exploration step first to understand our data better. In a previous section in this chapter, we carried out the data exploration and found that our data was unclean and had a lot of missing values. Next, in the code, we will replace the missing values in the features that we are planning to use to train our models. For training our models, we will use three features, namely, loan amount, applicant income, and credit history. For credit history, we replaced the missing value with the most occurring value out of the two values, that is, good and bad credit history and that turned out to be good as we showed in the previous section.

For applicant income, we find the average by firing a Spark SQL query, as shown next:

```
Dataset<Row> rowAvgDSIncome = spark.sql("select avg(_c6) avgIncome from loans");
rowAvgDSIncome.show();
```

Decision Trees

This would print the average value for applicant income as follows:

```
+-----------------+
|........avgIncome|
+-----------------+
|5403.459283387622|
+-----------------+
```

Thus, the rounded value for this is `5403`. Also, similar to this, we can find the average value for loan amount in the same way.

We will now pull out the row that contains this average value for the applicant's income:

```
Row avgIncRow = rowAvgDSIncome.collectAsList().get(0);
```

Now, we will pull the data from the dataset and map it to our JavaBean object:

```
JavaRDD<String> rowRdd = 
    spark.sparkContext().textFile("data/loan/loan_train.csv",
1).toJavaRDD();-
```

After the data is loaded in the `Rdd` object, we next map it our POJO object `LoanVO`, and while doing so, we also put in the missing values for our features:

```
JavaRDD<LoanVO> loansRdd = rowRdd.map( row -> {
        String[] dataArr = row.split(",");
        LoanVO lvo = new LoanVO();

        if(null == dataArr[6] || "".equals(dataArr[6]))
          lvo.setApplicantIncome(avgRow.getDouble(0));
        else lvo.setApplicantIncome(Double.parseDouble(dataArr[6]));


        if(null == dataArr[8] || "".equals(dataArr[8]))
          lvo.setLoanAmount(avgRow.getDouble(0));
        else lvo.setLoanAmount(Double.parseDouble(dataArr[8]));

        if(null == dataArr[10] || "".equals(dataArr[10]))
          lvo.setCreditHistory(avgCreditHistory);
        else lvo.setCreditHistory(Double.parseDouble(dataArr[10]));

        if(dataArr[12].equals("Y")) lvo.setLoanStatus(1.0);
        else lvo.setLoanStatus(0.0);

        return lvo;

});
```

Now we will create our **dataset** object. As we are using the new Spark ML API throughout this book, we will be using the dataset object for training our machine learning models.

```
Dataset<Row> dataDS = spark.createDataFrame(loansRdd.rdd(), LoanVO.
class);
```

After fetching our dataset, we create a `StringIndexer` object by passing the result label to it. It's a simple class which takes the distinct value from our response variable and puts it in a string array. This value is then put in the output column that we specify on our `labelIndexer` object, and in our case, it is the result column as shown next:

```
StringIndexerModel labelIndexer = new StringIndexer()
            .setInputCol("loanStatus")
            .setOutputCol("result")
            .fit(dataDS);
```

In the next step, we vectorize our features, that is, we store the value of all the features that we are using to train our models in the form of a vector. For this, we will use the `VectorAssember` class, which will take an array of feature names in a column and spit out a vector (containing the features) in a new column on the dataset.

```
String[] featuresArr = {"loanAmount","applicantIncome","creditHisto
ry"};
VectorAssembler va =
new VectorAssembler().setInputCols(featuresArr).
setOutputCol("features");
va.transform(dataDS);
```

Now we will split our dataset into two parts, one for training our model and the other for testing our model for its accuracy. We choose 70% of our data for our training and the remaining 30% for testing our trained model.

```
Dataset<Row>[] splits = dataDS.randomSplit(new double[]{0.7, 0.3});
Dataset<Row> trainingData = splits[0];
Dataset<Row> testData = splits[1];
```

Decision Trees

Create an object of the classifier model and provide it with the column that it should use to read its features for training, and also the column for the corresponding response variable. Also, as we mentioned earlier, we are using two kinds of split criteria with Apache Spark, and they are `entropy` and Gini Impurity. We specify the type of split criteria (that is, Entropy or Gini) that we want to use on our classifier by setting it as a property on our classifier, as shown next:

```
DecisionTreeClassifier dt = new DecisionTreeClassifier()
                .setLabelCol("result")
                .setFeaturesCol("features").setImpurity("entropy");
```

In the next step, we create an instance of an `IndexToString` object. Our model is mathematical, as such it produces mathematical output. To convert this output back to a string readable format, we use the `IndexToString` method. Similar to other classes that we have used from the Spark ML package, this class also takes an input column that it can read the input data from and an output column where it can produce the output result.

```
IndexToString labelConverter = new IndexToString()
                .setInputCol("prediction")
                .setOutputCol("predictedLabel")
                .setLabels(labelIndexer.labels());
```

> In the preceding code, we also pass the `labelIndexer` object. As you can see in a previous step, we used label indexer and this has converted our categorial output into a string array and put it in a column. The result that is generated by our model will be a number that will be an index in this array. From this index, the actual string value can then be pulled by our `IndexToString` class object.

The next step is the most important step, where we hook all these pieces together. As you saw in the preceding code, we built different features or objects for different parts in the workflow, it is using the pipeline API that we will hook all the pieces together. For this, we will pass all the individual workflows built earlier and push them into the pipeline using the `setStages` method, as shown next:

```
Pipeline pipeline = new Pipeline()
    .setStages(new PipelineStage[]{labelIndexer, va, dt, labelConverter});
```

> This pipeline API is an excellent new approach by the Spark ML package. With this, the main advantage is maintaining your large-scale machine learning code. Suppose you want to try a new model, then you just have to replace the `DecisionTreeClassifer` with a new model. Next, you just add that new model to the pipeline stages, and all the remaining pieces will remain as is with minimum impact as they are all loosely coupled.

Now, pass the training data to our pipeline model. What this would do is run the training data through all the stages that is:

- It will pull the distinct response variables using string indexer and put them in an array
- Next, the features are extracted from the dataset and put in a vector format
- Finally, the model is fed with these features
- In the last stage, the model prediction results that will be obtained on test data are mapped to the labels from the first stage and the result can be pulled from the output column of this stage.

The code for this is shown next:

```
PipelineModel model = pipeline.fit(trainingData);
```

In the next step, we pass our test data to the model and record its predicted results in a `predictions` object.

```
Dataset<Row> predictions = model.transform(testData);
```

Let's now print the first 10 rows from our predicted results and see how it looks. For this, we will select the first 10 rows from our `predictions` object generated earlier.

```
predictions.select("predictedLabel", "result", "features").show(10);
```

Decision Trees

And this will print the following result:

```
17/03/30 08:54:33 INFO CodeGenerator: Code gene
+--------------+------+--------------------+
|predictedLabel|result|            features|
+--------------+------+--------------------+
|           1.0|   0.0|    [98.0,210.0,1.0]|
|           0.0|   0.0|   [216.0,1025.0,1.0]|
|           1.0|   1.0|    [35.0,1442.0,1.0]|
|           0.0|   1.0|    [93.0,1800.0,0.0]|
|           1.0|   1.0|   [114.0,1853.0,1.0]|
|           1.0|   0.0|   [100.0,1928.0,1.0]|
|           1.0|   1.0|[146.412162162162...|
|           1.0|   0.0|   [113.0,2031.0,1.0]|
|           1.0|   0.0|   [101.0,2045.0,1.0]|
|           1.0|   0.0|    [88.0,2058.0,1.0]|
+--------------+------+--------------------+
only showing top 10 rows
```

Let's now try to run some stats on our predicted result. We mainly want to find out how accurate our model is. This will help us train it better. For this, we will use the `MulticlassClassificationEvaluator` object. To this, we pass our prediction and the original result and pass the metric we want it to find out. In our case, the cmetric we are interested in is accuracy:

```
MulticlassClassificationEvaluator evaluator = new
MulticlassClassificationEvaluator()
            .setLabelCol("result")
            .setPredictionCol("prediction")
            .setMetricName("accuracy");
```

Run this evaluator on the predicted results dataset as shown next and print the accuracy value and test results:

```
double accuracy = evaluator.evaluate(predictions);
System.out.println("Accuracy = " + accuracy);
System.out.println("Test Error = " + (1.0 - accuracy));
```

This will print the results as follows:

```
Accuracy = 0.7647058823529411
Test Error = 0.23529411764705888
```

As you can see, our model has a 76% accuracy.

We need to try different combinations of features or split criteria or train with more data in order to increase the efficiency of our models.

With this we come to an end to a discussion on decision trees. Even though we have covered a lot of ground on the concepts of decision trees, still there are other concepts that we have not covered, for example, the concept of tree pruning. We would encourage the users to read more on these topics on Wikipedia or other sources.

Summary

In this chapter, we covered a very important and popular algorithm in machine learning called as decision trees. A decision tree is very similar to a flowchart and is based on a set of rules. A decision tree algorithm learns from a dataset and builds a set of rules. Based on these rules, it splits the dataset into two (in the case of binary splits) or more parts. When a new data is fed in for predictions based on the attributes of the data, a particular path is taken and this follows along the full path of rules in the tree until a particular response is reached.

There are many ways in which we can split data in a decision tree. We explored two of the most common ways called Entropy and Gini Impurity. In either of these cases, the main criteria is to use the split mechanism, which makes the split set as homogeneous as possible. Both Entropy and Gini Impurity are mathematical formulas or approaches and as such the entire model works on numerical data.

In the next chapter, we will learn the very important concept of ensembling, where, instead of a single algorithm working on a problem, we use a set of algorithms to work on a problem. We will see how such an approach enhances the accuracy of our predictions.

8
Ensembling on Big Data

Have you used a Kinect while playing video games on Microsoft Xbox? It's so smooth how it detects your motion while you are playing games. It enables users to control and interact with their game without using any external device like a game controller. But how does it do that? How does the device detect the user's motion from the camera and predict the command that the motion suggested? Some users on different forums have claimed that a powerful random forest machine learning algorithm runs behind it and the link for the same is https://www.quora.com/Why-did-Microsoft-decide-to-use-Random-Forests-in-the-Kinect. Though I am myself not sure how true this claim is, this example at least demonstrates at what scale and level this powerful machine learning algorithm has the potential to be used. Random forests are perhaps one of the best machine learning algorithms because of the accuracy they bring in the predicted results and because of their implicit feature selection. They should be within the skillset of every machine learning programmer.

In this chapter, we will cover:

- The main concepts behind ensembling and what makes it so powerful. We will also cover the advantages and disadvantages of this approach.
- An introduction to and information on the concepts of random forest. We will also learn when we can use this algorithm and its specific advantages.
- A real-world use case of predicting loan defaults using random forest.
- The concept of gradient boosting, another important approach.
- Replacing the machine learning algorithm in our real-word use case example with the gradient boosting algorithm and running it on the dataset for predictions.

Before we get into the details of the random forest algorithm, we must first understand the concepts of ensembling.

Ensembling

Imagine that a group of friends are deciding which movie they want to see together. For this, they select their movie of choice from a set of, say, five or six movies. At the end, all their votes are collected and read. The movie with the maximum votes is picked and watched. What just happened is a real-life example of the ensembling approach. Basically, multiple entities act on a problem and give their selection out of a collection of discrete choices (in the case of a classification problem). The selection that was suggested by the maximum number of entities is chosen as the predicted choice.

This explanation was a general approach to ensembling. From the perspective of machine learning, it just means that multiple machine learning programs act on a problem that can be either of type classification or regression. The output from each machine learning algorithm is collected. The results from all the algorithms are then analyzed with different approaches like voting, averaging, or by using another machine learning algorithm, and finally out of the selected outcomes the best outcome is picked. Let's look into some of these approaches:

- **Voting**: This is a simple approach that we mentioned earlier. Multiple machine learning algorithms act on a task and give their output. The output from these algorithms is collected and each outcome is voted for. The outcome with the maximum number of votes is selected. This is also depicted in the following figure; as you can see in the figure, both the machine learning algorithms 1 and 2 choose the outcome 1, and hence its number of votes is 2, which is more than the votes on the other outcome and this outcome is selected:

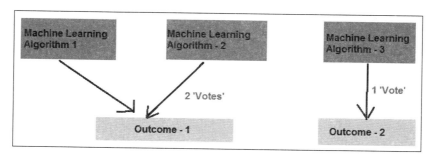

- **Averaging**: This is another simple approach but one that also yields good results. In this case, we also apply multiple algorithms on a problem and collect the output from each. The final outcome is the average of all the outcomes that were predicted by the machine learning algorithms. Suppose that you have a regression problem where you are predicting the age of a person based on different parameters:

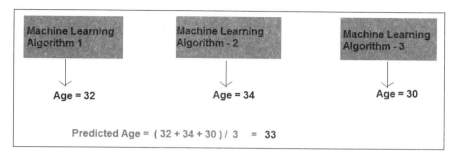

As seen in the previous image, the average age predicted by all the three algorithms is 33.

- **Using another machine learning algorithm**: This is another approach where the output of all the machine learning algorithms is fed to another machine learning algorithm and the final algorithm then picks the best possible algorithm. One example of this is an aritificial neural network, we will cover this in *Chapter 13, Deep Learning Using Big Data*.

What we just showed you are some of the approaches used in ensembling the models together. Next, we will cover some popular ensembling types.

Types of ensembling

Mentioned below are two common types of ensembling techniques that are used in random forests and gradient boosted trees algorithms in the big data stack:

Bagging

Before we understand bagging, we must look into a simple concept called **bootstrapping**.

Bootstrapping is a simple and powerful concept of pulling **n** samples of a fixed size from an existing training dataset. This is done until a decided number of samples is reached. Therefore, the existing training dataset now becomes a multi-training dataset, with each training set now containing **n** samples each. The approach is shown in the following diagram:

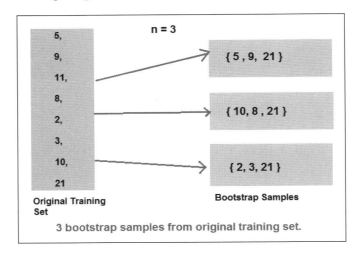

3 bootstrap samples from original training set.

As you can see in the previous diagram, three bootstrap samples of training datasets are created from the original training dataset. In practice, these sample datasets can run into thousands of datasets when this approach is used in the real world.

Now that we understand bootstrapping, bagging is a concept of training multiple machine learning models on these bootstrapped samples. So, on each bootstrapped sample a machine learning model is applied and predicts an outcome. Once all the outcomes of the machine learning models are generated, the final outcome is predicted using one of the approaches described previously, that is, voting, averaging, or by using another machine learning model. Bagging is depicted in the following figure:

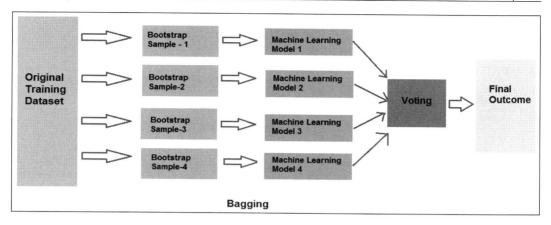

As you can see in the previous figure, the different machine learning models are trained on the sampling sets and the results of them are then voted for, with the final outcome selected based on the best voted result.

The approach of bagging that involves voting at the end of the result of multiple models is used in random forests for classification in Apache Spark. When averaging is used at the end for depicting the result of multiple models, this approach is used in regression using random forests in apache spark.

Boosting

Have you seen a relay race where members of the team take turns to run? The first runner runs first, followed by second and the third one, and so on. Similar to this sequential approach, boosting is the concept of running multiple weak-supervised learning algorithms one after the other, with the assumption of improving the results upon multiple runs.

The idea is that a weak machine learning algorithm might still be good on a certain part of the dataset for predictions, and by clubbing this strength of various weak algorithms we can come up with a strong predictive approach mechanism. Boosting is depicted in the following figure:

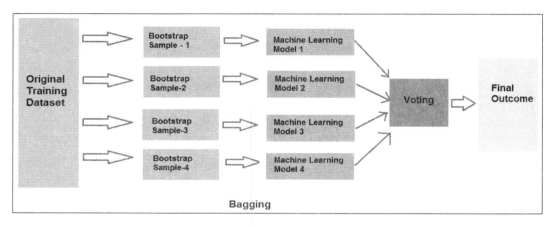

Bagging

The previous figure shows boosting. As we can see, there are multiple models that are used in the approach of boosting. There are some important points to note regarding the boosting approach:

- Each model is trained on the entire dataset
- The next model that is run after the first one has completed focuses on the errors made in predictions by the first model, and therefore it improves upon the results of the previous model, hence the name boosting

This approach of boosting is used in gradient boosted trees in apache spark. We have seen two popular techniques for improving the accuracy of your predictions. Let's now see some advantages and disadvantages of the ensembling approaches.

Advantages and disadvantages of ensembling

The advantages of ensembling are:

- The main advantage, and probably the only reason why ensembling is used so frequently, is that it improves the accuracy of the predictions. So, if you see and read the results of a lot of competitions on data analytics on sites like http://www.kaggle.com and so on, you would find that most of these competition winners would have used ensembling in some form.

- Ensembling can be used to represent complex relationships. Even though we are not showing you how we can combine two different types of models in this chapter, it is doable, and this form of combining different types of models in ensembling helps you to represent variable relationships using machine learning models.
- Another advantage of ensembling is it results in building a more generalized model with reduction in overfitting.
- As you will see when we discuss random forests in upcoming sections, the ensembling algorithms can be made to run in parallel across thousands of nodes. From the perspective of big data, random forest algorithms containing multiple machine learning algorithms can be easily distributed across a cluster of many distributed machines. Therefore, these algorithms scale well too.

We will now look at some of the disadvantages of using the ensembling approach:

- As you can see, this approach requires a training multiple model (which can run into hundreds, if not thousands, of models), and this approach can be tedious and slow. It is not suited for real-time applications that require immediate results.
- It makes the entire model prediction technical setup much more complex and hard to maintain as we have to take multiple components.

We have seen the approaches of ensembling, and their good and bad points, but how are these techniques used in the big data world? In the big data world, we have extensive Apache Spark usage and Apache Spark ships with some ensembled models within their machine learning APIs. We are going to learn two important models: random forest and **Gradient Boosted Trees (GBTs)**.

These ensembled models, that is, random forest and GBTs, are some of the best machine learning models used.

Random forests

In the previous chapter, we studied a very popular algorithm called decision trees. Recall that a decision tree is like a flow chart that splits the data using mathematical concepts like entropy before finally coming to a conclusion for predictive results. Like other machine learning algorithms decision trees suffer from the main problem of overfitting, *overfitting* as the name suggests is the problem when the model is not nicely generalized that is to say it gets too well trained on the training data such that it works well only on test samples from the training data but as soon as a new set of data is given to it (that which it did not see in training) it will perform very badly on that, so if our decision tree is made to classify a black sheep, it can classify a black dog as a black sheep due to overfitting. Throwing more entities at a problem is a common approach to solving problems. Hence, instead of one decision tree a cluster of decision trees can be used to solve the problem. Each tree can then classify or use regression to predict an outcome. The final outcome can then be decided on the basis of techniques like voting, averaging, and so on. The approach is shown in the following diagram:

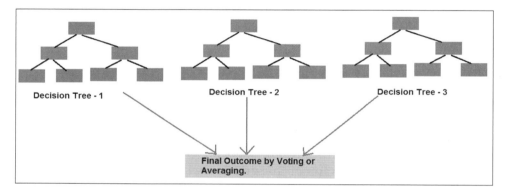

As seen previously, the decision trees feed their outcome to a voting or averaging system that then figures out the final outcome.

This approach of using multiple decision trees, or an ensemble of decision trees, is called **random forest**. From the perspective of big data, this approach is a clean fit on the big data stacks of Apache Spark and MapReduce. As you can see in the previous figure, the decision trees can be trained in parallel on a cluster of distributed machines and their results can then be collectively estimated.

As we saw in the advantages of ensembling, random trees main usage is to increase the accuracy of predictions. It also solves the main problem of overfitting. The trees in the forest can each be computed based on a different feature, which would enhance the performance by lowering the emphasis of a feature that is particularly noisy and overshadows other features.

Gradient boosted trees (GBTs)

Gradient boosted tree (**GBT**) is a very popular technique and is used extensively in many data science competitions online. You can check `kaggle.com` for this, where you will see that a lot of people have used it frequently. The idea behind this technique is that a group of weak learners can be boosted to produce a strong predictor. It is a sequential approach whereby in every stage of the sequence, a weak model is used to predict an outcome. The error of the previous model is boosted and the new models then give preference to these data items with error.

Let's try to understand the base concept of gradient boosting using an example. Suppose you are given a huge dataset containing data for users' age, their smoking habits, whether they smoke or not, and whether they are diabetic or not. Based on this data, you already have statistics for what their health insurance quotes are. Now our task is to train gradient boosted trees with this data and to come up with good predictions regarding the insurance quotes of the users.

In this case, a simplistic approach for gradient boosted tree is needed, as shown in the following diagram:

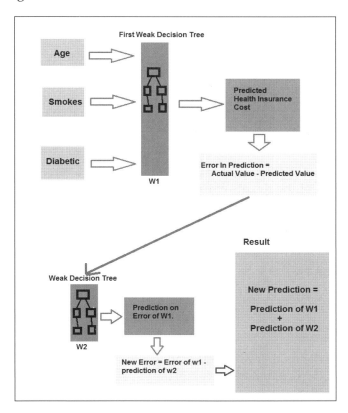

As seen in the previous diagram, the whole boosting technique goes as follows:

- First, a weak decision tree, **W1** (not a good performer), is trained on the dataset and is made to predict on the test set.
- The predictions are then evaluated, and the difference between the actual value and prediction is found out; this is called the residual. The predicted value can be depicted as follows:

$$y = F(x) + error$$

Here, **y** is the actual value and F(x) is the predicted value (which is a function of the feature x and *error* is the error in prediction). Error or residual can now be written as follows:

$$error = y - F(x)$$

- Next, we use this *error* and use another model (**W2**) on it to make a prediction. Thus, the predicted value of second model can now be depicted as:

$$error = Func2(X) + error2$$

Here, as you can see, the previous *error* is now equal to the function (*Func2*) of the second model and the *error2* is the error generated by the second model.

- Thus, we can now change our original function to a new function, which is a combination of the results of model 1, as well as the second model shown as follows:

Decision Tree 1	$y = F(x) + error$
Decision Tree 2	$error = Func2(X) + error2$
Final Result	$y = F(x) + Func2(X) + error2$

Thus, both the models are combined to produce a better result. *Error2* is lesser than the original error value, that is. *error*.

- Boosting is a technique that uses multiple models and an approach that is gradient descent, which minimizes the error going forward. Thus, the final predicted value will be the sum of the prediction of the multiple models as follows:

$$FinalPrediction = d_{tree1}(x) + d_{tree2}(x) + \cdots$$

 It's good to know about GBTs and gradient descent in depth. We would encourage readers to read more on this topic on Wikipedia and in other books on statistical learning.

So much for the theory, let's now dive into a real-world application usage of random forests and GBTs.

Classification problem and dataset used

We will be using random forests to predict a loan default by users. For this, we will be using a real-world dataset provided by Lending Club. Lending Club is a fintech firm that has publicly available data on its website. The data is helpful for analytical studies and it contains hundreds of features. Looking into all the features is out of the scope of this book. Therefore, we will only use a subset of features for our predictions. The features that we will be using are the following:

Feature	Description
ID	Unique numerical ID for each loan
Loan amount	Amount applied by the borrower for loan
Funded amount	The amount committed to that loan at that point in time
Annual income	Annual income of the applicant
Grade	Assigned loan grade
SubGrade	A more detailed loan grade; values are from A1 to G5
Home ownership	The home ownership status provided by the borrower; the values are RENT, OWN, MORTGAGE, and OTHER
Employment length	Length of time in current employment
Loan status	Status of loan, whether paid or other statuses
Zip code	Code of the state where the loan was applied for, for example, NJ for New Jersey, NY for New York

This dataset would give you a real taste of data in the real world, as this data is completely unclean and requires a lot of modifications before it can be used for analysis.

Before we train our models using this data, let's explore the data first for initial insights.

Data exploration

Before we do anything meaningful in the data exploration, we need to make sure our source of data that is our dataset files are present in HDFS where we expect to read them from our Spark jobs. To put the files in HDFS first bring the files to the operating system (that Linux in our case) and from Linux you can copy them to HDFS using the following command:

```
hdfs dfs -put <FILE_NAME><Destination directory in hdfs>
```

As seen we just use the `put` command to put a local file to HDFS.

 If you do not want to move this file to HDFS and just test it from your local OS like on Windows, you can do that too. For this you would have to load the file directly from Spark from your local filesystem.

For data exploration, we will first load the dataset using Apache Spark and see the number of rows in our dataset. For this, we will create `sparkContext` and initialize the `SparkSession`. To keep the code concise, we are not showing the full implementation as follows:

```
SparkConf c = ...

SparkSession spark = ...
```

Next, we load the dataset and find the number of rows in it:

```
JavaRDD<String> dataRdd =
spark.sparkContext().textFile("hdfs://datasets/LoanStats3a.csv",1)
.toJavaRDD();
```

As you can see previously, we loaded the dataset from an HDFS location and stored it in an RDD of strings. Luckily, this dataset is relatively clean and has one row per data item. Now to count the number of rows in the dataset we run the following:

```
System.out.println("Number of rows -->" + dataRdd.count());
```

This would print the result as follows:

```
Number of rows -->42538'
```

We are not interested in all the rows of this data, especially the current loans (as we are learning from historical data), and so we filter them out:

```
JavaRDD<String> filteredLoans = dataRdd.filter( row -> {
  return !row.contains("Current");
});
```

Also, there are lots of rows that contain N/A fields with no data, so we will filter them out, too. After filtering, we will load the data in POJOs and store them in the rdd object.

```
JavaRDD<LoanVO>data = rdd.map( r -> {
     if(r.size() < 100) returnnull;
     LoanVO lvo = new LoanVO();
     String loanId = r.getString(0).trim();
     String loanAmt = r.getString(2).trim();
     String fundedAmt = r.get(3).toString().trim();
     String grade = r.get(8).toString().trim();
     String subGrade = r.get(9).toString().trim();
     String empLength = r.get(11).toString().trim();
     String homeOwn = r.get(12).toString().trim();
     String annualInc = r.getString(13);
     String loanStatus = r.get(16).toString().trim();

     if(null == annualInc || "".equals(annualInc) ||
        null == loanAmt || "".equals(loanAmt) ||
        null == grade || "".equals(grade) ||
        null == subGrade || "".equals(subGrade) ||
        null == empLength || "".equals(empLength) ||
        null == homeOwn || "".equals(homeOwn) ||
        null == loanStatus || "".equals(loanStatus))
  returnnull;


    if(loanAmt.contains("N/A") ||
loanId.contains("N/A") || fundedAmt.contains("N/A") ||
grade.contains("N/A") ||
     subGrade.contains("N/A") || empLength.contains("N/A") ||
homeOwn.contains("N/A") || annualInc.contains("N/A") ||
loanStatus.contains("N/A")) returnnull;
```

```
            if("Current".equalsIgnoreCase(loanStatus)) returnnull;

        lvo.setLoanAmt(Double.parseDouble(loanAmt));
    lvo.setLoanId(Integer.parseInt(loanId));          lvo.
setFundedAmt(Double.parseDouble(fundedAmt));
    lvo.setGrade(grade);
    lvo.setSubGrade(subGrade);
    lvo.setEmpLengthStr(empLength);
    lvo.setHomeOwnership(homeOwn);
lvo.setAnnualInc(Double.parseDouble(annualInc.trim()));
    lvo.setLoanStatusStr(loanStatus);

    if(loanStatus.contains("Fully")) lvo.setLoanStatus(1.0);
    elselvo.setLoanStatus(0.0);
    returnlvo;

} ).filter(f ->  {
    if(f == null) returnfalse;
    elsereturntrue;
});
```

In the previous code, there are few notable things.

When the string contained N/A, we returned null and all null mapped values are filtered out in the RDD. For the loan status, we only picked loans that were fully paid and consider others in bad status. This is a slightly extreme guess, but for the first version of our program it's okay.

Next, we build the DataFrame out of this `rdd` and register it as a view to fire queries on:

```
Dataset<Row> dataDS = spark.createDataFrame(data.rdd(), LoanVO.class);
dataDS.createOrReplaceTempView("loans");
```

Now our data is ready to fire queries on. First, we will figure out the average, minimum, and maximum values of the loan amount, as this would give us a general idea of within what range people apply for loans on Lending Club. Apache Spark SQL provides us with a handy `describe` method that will help us calculate these values:

```
dataDS.describe("loanAmt").show();
```

This would generate the output as:

```
17/04/25 08:52:14 INFO DAGScheduler
+-------+------------------+
|summary|           loanAmt|
+-------+------------------+
|  count|             41597|
|   mean|10993.917229607905|
| stddev| 7358.233229929296|
|    min|             500.0|
|    max|           35000.0|
+-------+------------------+
```

As you can see, the max loan amount applied for is 35000$ (since this is in dollars) and the average loan amount applied for is around 11000$ (or 10994$ precisely).

Similar to this, let's see the average value for annual income:

```
dataDS.describe("annualInc").show();
```

This would print the following output. As you can see, the average annual income is close to 69000$ and the maximum is 6 million dollars:

```
+-------+------------------+
|summary|         annualInc|
+-------+------------------+
|  count|             42086|
|   mean| 69071.28295846605|
| stddev|64280.031646157455|
|    min|            1896.0|
|    max|         6000000.0|
+-------+------------------+
```

Let's look at the average of one more entity, that is, the funded amount, before we find a relation between these entities. This is the actual amount of loan that is approved and funded. As you can see in the following code, the average loan given is around 11000$ and the maximum given is 35000$.

```
+-------+------------------+
|summary|         fundedAmt|
+-------+------------------+
|  count|             42086|
|   mean| 10746.7673335551|
| stddev|7094.578580894198|
|    min|             500.0|
|    max|           35000.0|
+-------+------------------+
```

We have seen the general statistics for three important parameters, so let's try to plot on a scatter plot the values of the funded amount versus the annual income. To build the chart, we first load our dataset, clean it and extract the data, and fill it into an `rdd` containing Java POJOs (as was shown previously in this section). Next, we extract the two attributes, annual income and funded amount, and put these values in a chart object that is specific to the JFreeChart library used as follows:

```
JavaRDD<Double[]> dataRowArr = data.map(r -> {
Double[] arr = new Double[2];
  arr[0] = r.getAnnualInc();
    arr[1] = r.getFundedAmt();
    return arr;
});
```

Next, we use this RDD, collect it, and fill the data into *x* and *y* coordinates for the `XYSeries` object of JFreeChart. Finally, these coordinates are populated in the `XYSeriesCollection` dataset and returned.

```
List<Double[]> dataItems = dataRowArr.collect();
XYSeriesCollection dataset =new XYSeriesCollection() ;
XYSeries series= new XYSeries("Funded Loan Amount vs Annual Income of
       applicant");
   for (Double[] darr : dataItems) {
      Double annualInc = darr[0];
      Double fundedAmt = darr[1];
      series.add(annualInc, fundedAmt);
   }
     dataset.addSeries(series);
     return dataset;
```

To maintain conciseness, we are not showing you the full code here, but the full code can be found on our GitHub code repository. You can also refer to *Chapter 3, Data Visualization*, where we covered the code for building charts in detail.

This would print the following chart:

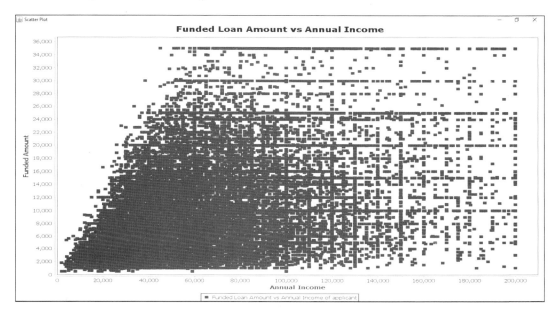

As you can see in the previous graph, and as expected, when the annual income increases the funded amount also generally increases.

> To draw a better graph with fewer outliers we have only taken the income range up to 200k above. This is to consider the fact that only a low percentage will have income above that range. As you can see in the graph, most people have an income of less than 100k.

There is a column for grade in the dataset. The grading of loans is an important concept as we believe that it directly impacts the loan amount funded. To check the value of grades we will check the number of loans that are bad loans, that is, they are not fully paid, and then check their grades on a bar chart.

Making a bar chart with JFreeChart is simple. From our dataset that we collected earlier, we will select loans with a bad status and check their count by firing a SQL query, as follows:

```
Dataset<Row> loanAmtDS = spark.sql("select grade,count(*) from loans
where
loanStatus = 0.0 group by grade");
```

Next, we collect data from this dataset and fill this data into a `DefaultCategoryDataset` of JFreeChart and return. This JFreeChart dataset object is then used to populate a bar chart component. To maintain brevity, we are not showing all of the code here.

```
final DefaultCategoryDataset dataset = new DefaultCategoryDataset( );
List<Row> results = loanAmtDS.collectAsList();
for (Row row : results) {
  String key = row.getString(0);
    dataset.addValue( row.getLong(1) , category , key );
}
return dataset;
```

This would print the chart as follows:

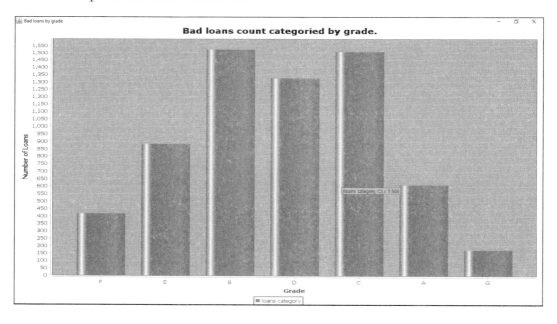

As you can see in the previous chart, most loans in the categories **B**, **C**, and **D** have defaulted or are bad loans.

Finally, we will see one more piece of data exploration before diving into predictive analytics on this dataset. We will see which zip code resulted in the maximum number of default loans, and we will pick the top ten such zip code and plot them on a bar chart. The code is very similar to the bar chart we drew previously for bad loan counts by grade. The query to pull the defaulted loan count by zip code will change, as follows:

```
Dataset<Row> loanAmtDS = spark.sql("select zipCode,count(*) loanCount
from loans
where loanStatus = 0.0 group by zipCode order
by loanCount desc limit 10");
```

This would print the chart as follows:

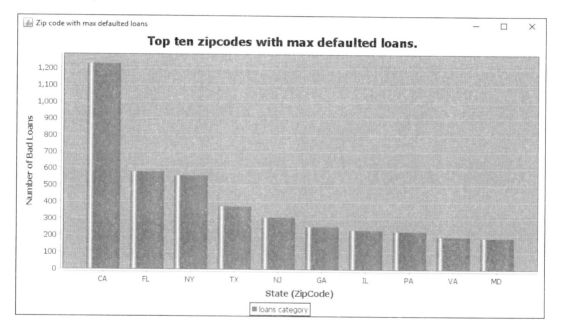

As you can see in the previous chart, California has the maximum number of bad loans, followed by Florida and New York, and so on.

> The bar chart for `zipCode` shown previously just shows the number of bad loans issued by different US states from Lending Club, but it does not show the overall performance of defaulted loans by state. To figure that out, you need to find the ratio of loans approved versus the loans defaulted.

We have just scratched the surface of data exploration. There is plenty of more stuff that you can do here, for example, you can use the remaining features that are present in this dataset for further exploration. We leave it to the readers to practice more of this stuff on their own.

Let's now finally get into the details of training our random forest model on this dataset.

Training and testing our random forest model

Before we train our model, we need to create the boiler plate code for creating the `SparkSession` object. For conciseness, we are just showing you the instance name as follows, for the full code, refer to our GitHub repository:

```
SparkSession spark = ...
```

Next, we load the dataset from HDFS (as this is a big data analysis case study, the data would be stored in HDFS or some other big data filesystem). Since the data is in the CSV format and contains double quotes (with double quotes containing commas internally, too), it's better to use the Spark CSV inbuilt package to load this data. If you want to extract this CSV without using any package then you would require cleaning of data, as this data contains special characters within the quotes:

```
Dataset<Row> defaultData = spark.read().csv("hdfs://datasets/LoanStats3a.csv");
```

This would load the CSV data into your `defaultData` dataset object. Next, we convert it back to `rdd` and run a `mapper` function on that `rdd`. This is done to extract data from this `rdd` and populate it into a Java POJO object. It is here we clean our data as well. To maintain brevity; we are not showing you the full code for this mapper:

```
JavaRDD<Row> rdd = defaultData.toJavaRDD();
JavaRDD<LoanVO> data = rdd.map( r -> {
if(r.size() < 100) return null;
LoanVO lvo = new LoanVO();
  String loanId = r.getString(0).trim();
  String loanAmt = r.getString(2).trim();
  ...
```

As you can see previously, we convert our dataset to `rdd` first and then invoke the `mapper` function on it. Within this `lambda` function, we are first checking whether the column count is good or not. If the column count is less that means we have missing data, but to keep things simple we will simply discard it by returning null. Finally, we build an instance of `LoanVO` POJO and also instantiate variables where we store extracted data.

Next, we run three rules on this data for cleaning it, as follows:

```
if(null == annualInc || "".equals(annualInc) ||
null == loanAmt || "".equals(loanAmt) || ... ) return null;

if(loanAmt.contains("N/A") || loanId.contains("N/A") || ... )
return null;

if("Current".equalsIgnoreCase(loanStatus)) return null;
```

First we remove the variables that have null values. If we locate any null value in our features we discard that row. Next, we check whether the N/A field is present in the data; this dataset contains some such fields. Even those fields are considered useless for our training and we discard that row, too. Finally, we check the status of the loans; if the loan status is current that means the loan is still in process and it is neither paid fully nor defaulted, and hence we ignore this data.

Next, we start setting the values of the extracted data into our POJO object:

```
lvo.setLoanAmt(Double.parseDouble(loanAmt));        lvo.
setLoanId(Integer.parseInt(loanId));
    ...
lvo.setLoanStatusStr(loanStatus);

if(loanStatus.contains("Fully")) lvo.setLoanStatus(1.0);
else lvo.setLoanStatus(0.0);
```

As you can see, we trade the loan status field in a special way for our models - we use numerical data for the loan status, and hence for a fully paid loan we put 1.0 and for the remaining ones we put 0.0. Finally, we return the POJO object that is filled with the extracted data. Along with the map method we also have a filter method that will filter out any null data that was returned; this method is useful for ignoring data that was ignored due to the specific rules described previously:

```
return lvo;
} ).filter(f -> {
if(f == null) return false;
 else return true;
});
```

As you saw in this example, we mostly discarded bad or missing data to keep things simple. You can also use techniques to fix the data such as replacing it with mean values or nearest values.

Now we convert our `rdd` to a dataset as we are using the Spark ML API, which works on dataset objects:

```
Dataset<Row> dataDS = spark.createDataFrame(data.rdd(), LoanVO.class);
dataDS.createOrReplaceTempView("loans");
```

Next, we will index the given results and store them in a new column using the `StringIndexer` class provided by Apache Spark. As the API shows we just need to provide the input column where the indexer can read the data, and the output column where it can generate the full output after running or fitting on the entire dataset:

```
StringIndexerModel labelIndexer = new StringIndexer()
        .setInputCol("loanStatus")
        .setOutputCol("indexedLabel")
        .fit(dataDS);
```

This dataset has lots of features as it is a real-word dataset, but we are only using loan amount, annual income, home ownership, funded amount, and grade for training our model. Recall that grade and home ownership are categorical fields with discrete string values as the output, but our models are mathematical and understand only numerical data. Therefore, we need to convert our categorical values to numerical values. To do this, we will use a handy `StringIndexerModel` provided by the Apache Spark features package:

```
StringIndexerModel gradeIndexer = new StringIndexer()
        .setInputCol("grade")
        .setOutputCol("gradeLabel")
        .fit(dataDS);

StringIndexerModel homeIndexer = new StringIndexer()
        .setInputCol("homeOwnership")
        .setOutputCol("homeOwnershipLabel")
        .fit(dataDS);
```

As you can see in the previous code, we provided the input and output columns and fitted this `StringIndexerModel` on the entire dataset. It would then convert our categorical values to numbers.

Just like other machine learning models, our random forest model also works on a vector of values. We need to convert our features data into vectors containing this data. We will use the `VectorAssembler` class from the Spark features package, and this class will convert our features into the vector form when provided with an array of our feature names. The feature names are nothing but the column names within our dataset:

```
String[] featuresArr =
{"loanAmt","annualInc","fundedAmt","gradeLabel","homeOwnershipLabel"};

VectorAssembler va = new
VectorAssembler().setInputCols(featuresArr).setOutputCol("features");
```

This `VectorAssembler` would pull the data from these columns and create a vector out of it. We need to now split out dataset into training and test dataset pieces. We will invoke the `randomSplit` method on our dataset and split the data. We are using 80% of the data for training and the remaining for testing:

```
Dataset<Row>[] splits = dataDS.randomSplit(new double[] {0.8, 0.2});
Dataset<Row> trainingData = splits[0];
Dataset<Row> testData = splits[1];
```

> 80/20 split for training and test data is usually a good starting point and is also known as *Pareto Principle*. We can always test with different training and test data variations. As you would do that, you would see that with more training data you can always train your models well but then you won't be able to see variations in your test results as the amount of test data is less now. So to match a perfect balance an 80/20 principle for dividing data is usually a good starting point.

Next, we build the instance of our classifier, in our case it is `RandomForestClassifier`. We provide the input column where the classifier can read the data and the output column where it can store its output. Since we are dealing with a random forest that will create multiple trees for us in an ensemble, we will also specify the number of trees to build as 100 (you can change this value as per your need):

```
RandomForestClassifier rf = new RandomForestClassifier()
        .setLabelCol("indexedLabel")
        .setFeaturesCol("features")
        .setImpurity("entropy").setNumTrees(100);
```

We specified the impurity or the split criterion as `entropy`; you can also use `gini` here.

We had earlier converted our output labels to indexes using a string indexer. Our results would be in the form of indexes too; we can convert them to proper strings using an index to string converter which also part of the Spark feature package. Again provide the input and output columns:

```
IndexToString labelConverter = new IndexToString()
        .setInputCol("prediction")
        .setOutputCol("predictedLabel")
        .setLabels(labelIndexer.labels());
```

Finally, we hook up all these workflow steps in our machine learning program using the pipeline API shown as follows. We provide all the instances of the objects that we created earlier to our `pipeline` object:

```
Pipeline pipeline = new Pipeline()
.setStages(new PipelineStage[] {labelIndexer,gradeIndexer,homeIndexer,
va, rf,
labelConverter});
```

Next we provide the training data to the pipeline workflow. This would run the training data through all the workflow steps provided to the pipeline earlier. This would also return an instance of `PipelineModel` object:

```
PipelineModel model = pipeline.fit(trainingData);
```

Let's finally make predictions on this model by running it on the test data:

```
Dataset<Row> predictions = model.transform(testData);
Predictions.show();
```

Let's see the first few rows of our predicted results:

```
+----------+----------------+--------------------+--------------------+--------------------+----------+--------------+
|gradeLabel|homeOwnershipLabel|            features|       rawPrediction|         probability|prediction|predictedLabel|
+----------+----------------+--------------------+--------------------+--------------------+----------+--------------+
|       0.0|             1.0| 2.0|[2500.0,8628.0,25...|[87.3542582782462...|[0.87354258278246...|       0.0|           1.0|
|       0.0|             3.0| 2.0|[3600.0,9600.0,36...|[76.2244409474016...|[0.76224440947401...|       0.0|           1.0|
|       0.0|             1.0| 0.0|[5700.0,11476.0,5...|[89.7924986989646...|[0.89792498698964...|       0.0|           1.0|
|       1.0|             0.0| 0.0|[2400.0,12000.0,2...|[82.9439876266355...|[0.82943987626635...|       0.0|           1.0|
|       0.0|             1.0| 2.0|[2700.0,12000.0,2...|[88.3787951654340...|[0.88378795165434...|       0.0|           1.0|
|       1.0|             1.0| 0.0|[5000.0,12000.0,5...|[89.7924986989646...|[0.89792498698964...|       0.0|           1.0|
|       0.0|             0.0| 2.0|[1950.0,12000.0,1...|[82.7028090041320...|[0.82702809004132...|       0.0|           1.0|
+----------+----------------+--------------------+--------------------+--------------------+----------+--------------+
```

Note the columns containing our features (in vector form) and the raw prediction which is nothing but a probability and finally the actual prediction which is a binary as **1** or **0**.

 From the perspective of big data we can always store our trained model to external storage and recreate it back to make prediction using this storage. We can also store the results we have just generated into an external storage for example, in Parquet format so that later we can further analyze and query the results.

We have seen now that the output result is also a dataset which can be queried for further analysis. Let's now pull some statistics from the output results and figure out how accurate our random forest model is. Apache Spark provides us with a handy-out-of-the-box class for this and it is called as `MulticlassClassificationEvaluator`, and we just need to provide it with the actual results, the predicted results and the metric we are looking for (in our case we are interested in figuring out accuracy hence our metric is Accuracy). Finally, we can invoke the `evaluate` method on this class and it will give us the result of this metric Accuracy:

```
MulticlassClassificationEvaluator evaluator = new
MulticlassClassificationEvaluator()
        .setLabelCol("indexedLabel")
        .setPredictionCol("prediction")
        .setMetricName("accuracy");
double accuracy = evaluator.evaluate(predictions);
    System.out.println("Accuracy = " + (100 * accuracy));
    System.out.println("Test Error = " + (1.0 - accuracy));
```

As seen in the preceding code we have printed the `Accuracy` and the `Test Error` and the results would be as follows:

```
Accuracy = 84.4386853575215
Test Error = 0.15561314642478496
```

Around 84% accurate, which is pretty bad for a machine learning approach. Our aim from this chapter is to give enough tools to the readers to get started on this ensembling approach and work on tuning these models for better results.

As we mentioned before Gradient boosted trees can give better performance though they cannot be as parallelly trained as a random forest. Let's try to see this in action.

Training and testing our gradient boosted tree model

The use of gradient boosted tree model in place of the random forest is easy. Pretty much all the code shown previously is same except that we replace the random forest model with the gradient boosted classifier and fit it into the same pipeline API as we covered in the previous section. The code for the same is shown next and in the code we create the instance of the gradient boosted classifier and provide it the column where it can read the features and a column where it can print the predicted results. We also provide the split criterion, that is, `entropy`:

```
GBTClassifier rf = new GBTClassifier()
    .setLabelCol("indexedLabel")
    .setFeaturesCol("features").setImpurity("entropy");
```

After running this model we again collect the predicted results and using the multi-classification evaluator we figure out the accuracy of this classifier. The code is again exactly similar to the previous section code. The result would be printed as:

```
Accuracy = 84.61628588166373
Test Error = 0.15383714118336267
```

Approximately 85%, as you can see the performance is slightly improved. The readers should tweak the model parameters further or change the features further to check this. Gradient Boosting is a very powerful approach and has been successfully deployed in many production systems. There is one more algorithm called as XGBoost which is even more popular or perhaps the most popular one. We urge the users to read more on the XGBoost algorithm.

> Even though we have covered Random forests and Gradient boosting by using classification techniques, these algorithms can very well be applied for regression as well.

Before we finally wrap this chapter, I would like to touch on one point regarding how we could deploy these models to a production system. For deploying machine learning models in production we generally train the model on historical data and store it in external storage. When needed we rebuild the model from the external system and run the test data for predictions on it. The models are run using the `spark-submit` command, in fact training and storage of the model are also run using the `spark-submit` job. We covered some of this in *Chapter 1, Big Data Analytics with Java*.

Summary

In this chapter, we learnt about a very popular approach called ensembling in machine learning. We learnt how a group of decision trees can be parallelly built, trained, and run on a dataset in the case of random forests. Finally, their results can be combined by techniques like voting for classification to figure out the best voted classification or averaging the results in case of regression. We also learnt how a group of weak decision tree learners or models can be sequentially trained one after the other with every step boosting the results of the previous model in the workflow by minimizing an error function using techniques such as gradient descent. We also saw how powerful these approaches are and saw their advantages over other simple approaches. We also ran the two ensembling approaches on a real-world dataset provided by Lending Club and analyzed the accuracy of our results.

In the next chapter, we will cover the concept of clustering using the k-means algorithm. We will also run a sample case study and see how a retail store can categorize their customers into important groups based on this approach.

9
Recommendation Systems

When you go to a bookstore to buy books, you have a particular book in mind generally, which you are interested in buying and you look for that particular book in the bookshelves. Usually, in the book store, the top selling books at that point in time are kept upfront and the remaining inventory is kept on the shelves arranged (sorted). A typical small bookstore can have say a few thousand books or maybe more. So, in short, the limit to which the physical products are available is right in front of you as a customer and you can pick and choose what you like at that moment. Also, physical stores keep top products in front as they are more sellable, but there is no way the products can be arranged according to the choice or preference of a customer coming to a physical store. However, this is not the case when you go to popular online e-commerce store such as Amazon or Walmart. There could be a million if not a billion products on Amazon when you go to buy stuff on it. In fact, the range of information available in the case of an e-commerce store is so much that there is a need of an elegant way to present the most useful information to the customer when he is browsing the website so as to make him do a purchase on the site. Putting the top products from a set of millions of products (as it was done in physical stores) is not an elegant way as the space on the webpage the user is browsing is limited. To solve this issue most online e-commerce stores use recommendation systems.

When it comes to machine learning, one of the first references of its extensive usage that comes to mind are the use of recommendation systems. As the name suggests, these are software systems that recommend useful entities to end users. These useful entities can be anything whether they are songs or movies, or even articles on a tutorial website. In this chapter, we will study the following details about recommendation systems:

- Concepts, use cases, and types of recommendation systems
- Content-based recommendation system concepts
- A simple content-based recommendation system case study

Recommendation Systems

- Collaborative recommendation systems concepts
- Exploring the MovieLens movie review datasets
- Recommending movies to users using collaborative filtering

Recommendation systems and their types

Before we dig deeper into the concepts of the recommendation system, let's see two real-world examples of recommendation engines that we might be using on a daily basis. The examples are shown in the following screenshots. The first screenshot is from Amazon.com, where we can see a section called **Customers who bought this also bought**, and the second screenshot will be from YouTube.com, where we are seeing a section called **Recommended**:

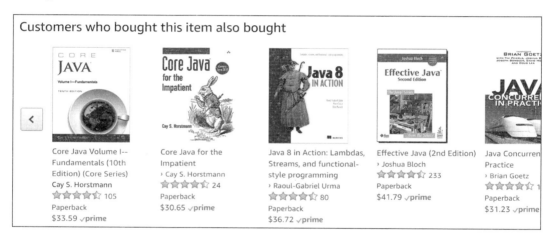

As you can see in the screenshot which we have taken from http://www.amazon.com, it shows a list of books on Java. So if you search for keyword Core Java on Amazon.com for buying books, you will get a list books on Core Java. If you select one of those core java books now and click on it, you will be directed to the page where you will get the full description about the book: its price, author, reviews, and so on. It is here, at the bottom of this section you will get a link as shown above where it is mentioned **Customers who bought this item also bought**. So Amazon.com here is giving suggestions or **recommendations** to its customers so that they can also simultaneously look into other items along with the one they are interested in purchasing.

Now let's see an image from another famous website www.youtube.com that frequently uses the concept of making recommendations to its users.

As you can see in the image Youtube.com is showing recommended videos to its users and these videos for recommendations are figured out on the basis of the taste or liking of their users (based on what they watched in the past).

Both these websites Amazon and YouTube contain billions of items. They have both figured out a unique and useful way of displaying items to their customers based on their taste. So if you go to www.youtube.com, you can see some suggestions of videos (under the **Recommendations** section in the image) you might like on the right side of the screen and similar to this when you are browsing some product on Amazon you also get suggestions of other products that you might like to buy. This is a very useful technique given that these websites have millions if not billions of products then how do they figure out which products or videos the users would like. This is all what recommendation systems are all about. Recommendation systems are everywhere these days whether in e-commerce where they display the products the users might be interested in, or in music sites where they show you the next song you might like to play or they create an auto-playing playlist based on your taste, for example, as in Pandora or on fashion-related websites they might show you the next piece of fashionable items that might appeal to you. In fact, their use is so much that it is said that on a giant e-commerce store the recommendation systems are responsible for driving a lot of their sale.

> Consider this when you go to Amazon.com to buy a book and you are browsing it, you might see a similar book that was published may be fifty years earlier. This old book might turn out to be more appealing to you and you might buy it. The point here is that Amazon.com uses recommendation systems power to pull this old book from their inventory and show it to you. However, the same thing is just not possible on a physical store where the inventory is limited and there is no way to connect items that might belong to the same taste.

As the name suggests, the recommendation systems gives recommendations or suggestions to the users regarding products they might like. There are various ways which is figured out either based on the click history of the users or based on the ratings they gave to other products on these websites. In all cases, some form of machine learning techniques are used to predict based on historical data of user's browse history or product transactions.

There are two types of recommendation systems, namely:

- Content-based recommendation systems
- Collaborative recommendation systems

We will cover both in detail in the upcoming sections.

Content-based recommendation systems

In content-based recommendations, the recommendation systems check for similarity between the items based on their attributes or content and then propose those items to the end users. For example, if there is a movie and the recommendation system has to show similar movies to the users, then it might check for the attributes of the movie such as the director name, the actors in the movie, the genre of the movie, and so on or if there is a news website and the recommendation system has to show similar news then it might check for the presence of certain words within the news articles to build the similarity criteria. As such the recommendations are based on actual content whether in the form of tags, metadata, or content from the item itself (as in the case of news articles).

Let's try to understand content-based recommendation using the following diagram:

As you can see in the preceding diagram, there are four movies each with a specific director and genre. Now, look at **Movie - 1** and **Movie - 3**. Both these movies have the same director as well as genre, thus they have a very strong similarity. Thereby, when somebody sees **Movie - 1**, then **Movie - 3** can be recommended to him. Now, look at **Movie - 4** which has a director that is different from **Movie - 1** and **Movie - 3** but its genre is similar to **Movie - 1** and **Movie - 3**. Thereby **Movie - 4** is more similar to **Movie - 1** and **Movie - 3** as compared to **Movie - 2** (because **Movie - 2** has a totally different director as well as a different genre), and hence it can be the second recommendation to **Movie - 1** after **Movie - 3**. This is all what content recommendation is in simple terms.

Figuring similarity by looking at the preceding diagram is easy for human beings, but how does a computer figure this out. To figure out similarity between items using their attributes, simple mathematical technique or formulas are used, such as cosine similarity, Euclidean distance, Pearson coefficient, and so on. From the perspective of machine learning models, we need to create a vector of features and then find the distance between these vectors using these mathematical formulas.

Recommendation Systems

Let's try to understand the concept of finding similarity using an example. Let's suppose we have three movies with the following two properties, that is, their average rating by the viewers and the rating they received on how much action is there in the movie:

Star Wars	Average Rating = 5, Action = 5
Independence Day	Average Rating = 4, Action = 4
Love at First Sight	Average Rating = 5, Action = 1

Now, our job is to find how similar are these movies to each other. Using the properties or attributes some mathematical formulas can be applied that could give us the similarity value. Some of these formulas are as follows:

- **Euclidean Distance**: This is a very simple concept of plotting the properties on a chart. In the preceding example, we have two properties only, so we can plot them on a two-dimensional chart on x/y axis as shown in the following diagram:

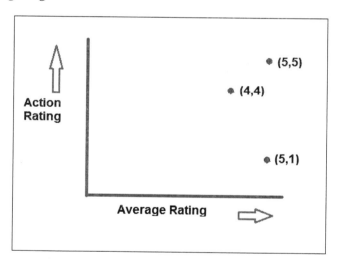

As you can see, we have plotted the images on the x and y axis with x being the average rating and y being the action rating and the x and y coordinates are also written next to the points. Now, using the naked eye, which points do you think are closer together? As you can see, point *(4,4)* or movie Independence Day is closer to point *(5,5)* which is Star Wars. This is not a bad prediction as anybody who has seen these movies will know that both are action-packed awesome movies.

To calculate this distance between the points, you can use the Euclidean formula as follows:

$$Euclidean\ Distance\ between\ points\ (x1, y1)\ and\ (x2, y2)$$
$$= \sqrt{(x1-x2)^2 + (y1-y2)^2}$$

Thus, the distance between Star Wars and Independence Day is =

$$\sqrt{(5-4)^2 + (5-4)^2} = 1.4$$

And the distance between Star Wars and Love at First Sight is =

$$\sqrt{(5-5)^2 + (5-1)^2} = 4$$

As the distance, between Star Wars and independence day is 1.4, which is less than the other distance (that is, 4), these movies are closer or more similar.

> Even though we have used only two properties here but the formula can be used on *n* number of properties in the same way. Just keep adding more and more properties within the square root function.

Recommendation Systems

- **Pearson Correlation**: This is one of the most popular methods of finding similarity between two datapoints and is heavily used in real applications. This measure tries to find the linear relationship between two datapoints as such it's not a good measure to check similarity between nonlinear datapoints. What it tries to do is to figure out the best fit line between the datapoints. If the line hits all the datapoints that means the entities represented by these datapoints are very similar, else they are less similar. It gives better results than the Euclidean method as in the case of Euclidean method the weight of the user's rating is not taken into account. That is to say that some users might be rating their movies very highly when they like them while other users even when they like some movie they might not give that high a rating. This might lead to movies getting bad scores even when the viewers liked them. In the case of Euclidean method, this would result in a bad distance calculations as Euclidean Distance just cares for the value of distance into account and does not take into account the **tendency** of scoring or rating by the users (that is to say, for example, some people might have a tendency to always give a score above 3, so the worst movie they watched might have a score of 3). Pearson's Correlation coefficient is not affected by this issue as it cares about the best fit line and not about the actual distance between the data points themselves.

Let's try to understand this using the same small dataset that we used in the Euclidean Distance, that is, for the movies **Star Wars**, **Independence Day**, and **Love at First Sight**. We will plot the **Star Wars** on the x axis and corresponding other movies on the y axis. Next, we will plot the average rating and action value for each movie, so for **Star Wars** and **Independence Day** the average rating coordinates will be (5,4) and (5,4) on the (x,y) axis. The charts for **Star Wars** versus both the movies is shown in the following diagram:

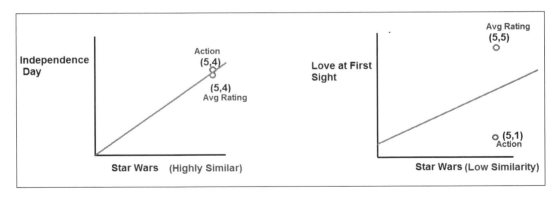

As you can see in the preceding charts for the **Independence Day** and **Star Wars** movies, the best fit line collides with the datapoints and they are very similar while in the case of **Star Wars** versus **Love at First Sight** the best fit line is far away from the datapoints and hence they are much less similar.

The exact value of similarity can be calculated using the Pearson's correlation mathematical formula as follows:

$$r = \frac{n\left(\sum xy\right) - \left(\sum x\right)\left(\sum y\right)}{\sqrt{\left[n\sum x^2 - \left(\sum x\right)^2\right]\left[n\sum y^2 - \left(\sum y\right)^2\right]}}$$

Here *r* is the Pearson Correlation value and *x* represents a dataset of *n* values as (x1, x2, x3…..) and *y* represents dataset of *n* values as (y1,y2,y3……). So if you want to find the Pearson Coefficient between say **Star Wars** and **Love at First Sight**, the values (or vectors) will be (average rating, action) for each movie, that is,(5,5) and (5,1). You might have observed that this is not a simple formula, but since this is being very popular, it is part of almost all the popular statistic libraries and is also available in the Spark RDD statistic module (in Spark MLlib).

Other than these two, there are more mathematical approaches for finding similarity such as Cosine Similarity, Jaccard Coefficient, and so on. We would encourage the users to read more about them on the Net or in other machine learning books.

Now, we have seen how a content-based recommendation system can be built using these mathematical formulas, but if you see closely, this approach requires you to compare each movie against each other and find the distance. For small amount of data, this is fine, but in the case of big data with millions of data points these calculation amounts will explode and this would become a very slow approach. To solve this, we can use collaborative recommendations as we will see in the next section.

But still under what condition would you use a Content Recommender?

Recommendation Systems

The Content Recommender solves the **cold start** problem. A cold start problem is nothing but when there is not much historical transaction data to look at. As in the preceding case, we just compared the movies using their properties to each other but we did not have any transaction data (like the ratings on those movies given by the site customers). Content Recommender as such is very good for completely new websites that do not have much web traffic, so they do not have user's transactional data, yet they want to give recommendations to the users for better user interactions.

So much for the theory, let's dig into an actual code.

Dataset

For this chapter, we are using the very popular movielens dataset, which was collected by the GroupLens Research Project at the University of Minnesota. This dataset contains a list of movies that are rated by their customers. It can be downloaded from the site https:/grouplens.org/datasets/movielens. There are few files in this dataset, they are:

- u.item:This contains the information about the movies in the dataset. The attributes in this file are:

Movie Id	This is the ID of the movie
Movie title	This is the title of the movie
Video release date	This is the release date of the movie
Genre (isAction, isComedy, isHorror)	This is the genre of the movie. This is represented by multiple flats such as isAction, isComedy, isHorror, and so on and as such it is just a Boolean flag with 1 representing users like this genre and 0 representing user do not like this genre.

- u.data: This contains the data about the users rating for the movies that they have watched. As such there are three main attributes in this file and they are:

Userid	This is the ID of the user rating the movie
Item id	This is the ID of the movie
Rating	This is the rating given to the movie by the user

- **u.user**: This contains the demographic information about the users. The main attributes from this file are:

User id	This is the ID of the user rating the movie
Age	This is the age of the user
Gender	This is the rating given to the movie by the user
ZipCode	This is the zip code of the place of the user

As the data is pretty clean and we are only dealing with a few parameters, we are not doing any extensive data exploration in this chapter for this dataset. However, we urge the users to try and practice data exploration on this dataset and try creating bar charts using the ratings given, and find patterns such as top-rated movies or top-rated action movies, or top comedy movies, and so on.

Content-based recommender on MovieLens dataset

We will build a simple content-based recommender using Apache Spark SQL and the MovieLens dataset. For figuring out the similarity between movies, we will use the Euclidean Distance. This Euclidean Distance would be run using the genre properties, that is, action, comedy, horror, and so on of the movie items and also run on the average rating for each movie.

First is the usual boiler plate code to create the SparkSession object. For this we first build the Spark configuration object and provide the master which in our case is local since we are running this locally in our computer. Next using this Spark configuration object we build the SparkSession object and also provide the name of the application here and that is ContentBasedRecommender.

```
SparkConf sc = new SparkConf().setMaster("local");
SparkSession spark = SparkSession
            .builder()
            .config(sc)
            .appName("ContentBasedRecommender")
            .getOrCreate();
```

Recommendation Systems

Once the `SparkSession` is created, load the rating data from the `u.data` file. We are using the Spark RDD API for this, the user can feel free to change this code and use the dataset API if they prefer to use that. On this Java RDD of the data, we invoke a `map` function and within the `map` function code we extract the data from the dataset and populate the data into a Java POJO called `RatingVO`:

```
JavaRDD<RatingVO> ratingsRDD =
spark.read().textFile("data/movie/u.data").javaRDD()
    .map(row -> {
                RatingVO rvo = RatingVO.parseRating(row);
                return rvo;
    });
```

As you can see, the data from each row of the `u.data` dataset is passed to the `RatingVO.parseRating` method in the `lambda` function. This `parseRating` method is declared in the POJO `RatingVO`. Let's look at the `RatingVO` POJO first. For maintaining brevity of the code, we are just showing the first few lines of the POJO here:

```
public class RatingVO implements Serializable {
    private int userId;
    private int movieId;
private float rating;
private long timestamp;
    private int like;
```

As you can see in the preceding code we are storing `movieId`, `rating` given by the user and the `userId` attribute in this POJO. This class `RatingVO` contains one useful method `parseRating` as shown next. In this method, we parse the actual data row (that is tab separated) and extract the values of rating, `movieId`, and so on from it. In the method, do note the `if...else` block, it is here that we check whether the rating given is greater than three. If it is, then we take it as a 1 or like and populate it in the `like` attribute of `RatingVO`, else we mark it as 0 or dislike for the user:

```
        public static RatingVO parseRating(String str) {
           String[] fields = str.split("\t");
...
           int userId = Integer.parseInt(fields[0]);
           int movieId = Integer.parseInt(fields[1]);
           float rating = Float.parseFloat(fields[2]);
if(rating > 3) return new RatingVO(userId, movieId, rating, timestamp,1);
           return new RatingVO(userId, movieId, rating, timestamp,0);
}
```

Once we have the RDD object built with the POJO's `RatingVO` contained per row, we are now ready to build our beautiful dataset object. We will use the `createDataFrame` method on the `spark` session and provide the RDD containing data and the corresponding POJO class, that is, `RatingVO`. We also register this dataset as a temporary view called `ratings` so that we can fire SQL queries on it:

```
Dataset<Row> ratings = spark.createDataFrame(ratingsRDD, RatingVO.class);
ratings.createOrReplaceTempView("ratings");
 ratings.show();
```

The `show` method in the preceding code would print the first few rows of this dataset as shown next:

```
17/05/05 08:50:54 INFO CodeGenerator: Code generat
+----+-------+------+---------+------+
|like|movieId|rating|timestamp|userId|
+----+-------+------+---------+------+
|   1|     50|   5.0|881250949|     0|
|   1|    172|   5.0|881250949|     0|
|   0|    133|   1.0|881250949|     0|
|   0|    242|   3.0|881250949|   196|
|   0|    302|   3.0|891717742|   186|
|   0|    377|   1.0|878887116|    22|
|   0|     51|   2.0|880606923|   244|
|   0|    346|   1.0|886397596|   166|
```

Next we will fire a Spark SQL query on this view (`ratings`) and in this SQL query we will find the average rating in this dataset for each movie. For this, we will group by on the `movieId` in this query and invoke an `average` function on the `rating` column as shown here:

```
Dataset<Row> moviesLikeCntDS = spark.sql("select movieId,avg(rating) likesCount from ratings group by movieId");
```

The `moviesLikeCntDS` dataset now contains the results of our group by query. Next we load the data for the movies from the `u.item` data file. As we did for users, we store this data for movies in a `MovieVO` POJO. This `MovieVO` POJO contains the data for the movies such as the `MovieId`, `MovieTitle` and it also stores the information about the movie genre such as action, comedy, animation, and so on.

The genre information is stored as 1 or 0. For maintaining the brevity of the code, we are not showing the full code of the `lambda` function here:

```
JavaRDD<MovieVO> movieRdd = spark.read().textFile("data/movie/u.
item").javaRDD()
                       .map(row -> {
                                    String[] strs = row.split("\\|");
                                    MovieVO mvo = new MovieVO();
                                    mvo.setMovieId(strs[0]);
                                    mvo.
setMovieTitle(strs[1]);
...
                                    mvo.setAction(strs[6]);
                                    mvo.setAdventure(strs[7]);
                                    ...
                                    return mvo;
});
```

As you can see in the preceding code we split the data row and from the splitted result, which is an array of strings, we extract our individual values and store in the `MovieVO` object. The results of this operation are stored in the `movieRdd` object, which is a Spark RDD.

Next, we convert this RDD into a Spark dataset. To do so, we invoke the Spark `createDataFrame` function and provide our `movieRdd`to it and also supply the `MovieVO` POJO here. After creating the dataset for the movie RDD, we perform an important step here of combining our movie dataset with the `moviesLikeCntDS` dataset we created earlier (recall that `moviesLikeCntDS` dataset contains our movie ID and the average rating for that review in this `movie` dataset). We also register this new dataset as a temporary view so that we can fire Spark SQL queries on it:

```
Dataset<Row> movieDS = spark.createDataFrame(movieRdd.rdd(),
MovieVO.class).join(moviesLikeCntDS, "movieId");
movieDS.createOrReplaceTempView("movies");
```

Before we move further on this program, we will print the results of this new dataset. We will invoke the show method on this new dataset:

```
movieDS.show();
```

This would print the results as (for brevity we are not showing the full columns in the screenshot):

```
-+--------------------+-------+-------+------------+-------+-----+--------+---+-------+------------------+
1|         movieTitle|musical|mystery|releaseDate|romance|sciFi|thriller|war|western|     averageRating|
-+--------------------+-------+-------+------------+-------+-----+--------+---+-------+------------------+
.|     Hoodlum (1997)|      0|      0| 22-Aug-1997|      0|    0|       0|  0|      0|2.9315068493150687|
.|Ice Storm, The (1...|      0|      0| 01-Jan-1997|      0|    0|       0|  0|      0|3.6436781609195403|
.|It's a Wonderful ...|      0|      0| 01-Jan-1946|      0|    0|       0|  0|      0| 4.121212121212121|
.|Heavenly Creature...|      0|      0| 01-Jan-1994|      0|    0|       1|  0|      0|3.6714285714285713|
.|Hunchback of Notr...|      1|      0| 21-Jun-1996|      0|    0|       0|  0|      0| 3.377952755905512|
.|American Presiden...|      0|      0| 01-Jan-1995|      1|    0|       0|  0|      0|3.6280487804878048|
.|       Congo (1995)|      0|      1| 01-Jan-1995|      0|    1|       0|  0|      0|2.4523809523809526|
.|Preacher's Wife, ...|      0|      0| 13-Dec-1996|      0|    0|       0|  0|      0| 2.926470588235294|
```

Now comes the turn for the meat of this content recommender program. Here we see the power of Spark SQL, we will now do a self join within a Spark SQL query to the temporary view movie. Here, we will make a combination of every movie to every other movie in the dataset except to itself. So if you have say three movies in the set as (movie1, movie2, movie3), this would result in the combinations (movie1, movie2), (movie1, movie3), (movie2, movie3), (movie2, movie1), (movie3, movie1), (movie3, movie2). You must have noticed by now that this query would produce duplicates as (movie1, movie2) is same as (movie2, movie1), so we will have to write separate code to remove those duplicates. But for now the code for fetching these combinations is shown as follows, for maintaining brevity we have not shown the full code:

```
Dataset<Row> movieDataDS =
spark.sql("select m.movieId movieId1,m.movieTitle movieTitle1,m.action action1,m.adventure adventure1, ... "
+ "m2.movieId movieId2,m.movieTitle movieTitle2,m2.action action2,m2.adventure adventure2, ..."
```

Recommendation Systems

If you invoke show on this dataset and print the result, it would print a lot of columns and their first few values. We will show some of the values next(note that we show values for movie1 on the left-hand side and the corresponding movie2 on the right-hand side):

```
+--------+--------------------+-------+----------     +--------------------+-------+----------+----------+-
|movieId1|         movieTitle1|action1|adventure1     |         movieTitle2|action2|adventure2|animation2|c
+--------+--------------------+-------+----------     +--------------------+-------+----------+----------+-
|     299|     Hoodlum (1997)|      0|         (     |Ice Storm, The (1...|      0|         0|         0|
|     299|     Hoodlum (1997)|      0|         (     |It's a Wonderful ...|      0|         0|         0|
|     299|     Hoodlum (1997)|      0|         (     |Heavenly Creature...|      0|         0|         0|
|     299|     Hoodlum (1997)|      0|         (     |Hunchback of Notr...|      0|         0|         1|
|     299|     Hoodlum (1997)|      0|         (     |American Presiden...|      0|         0|         0|
|     299|     Hoodlum (1997)|      0|         (     |        Congo (1995)|      1|         1|         0|
|     299|     Hoodlum (1997)|      0|         (     |Preacher's Wife, ...|      0|         0|         0|
|     299|     Hoodlum (1997)|      0|         (     |Associate, The (1...|      0|         0|         0|
|     305|Ice Storm, The (1...|      0|         (     |      Hoodlum (1997)|      0|         0|         0|
|     305|Ice Storm, The (1...|      0|         (     |It's a Wonderful ...|      0|         0|         0|
|     305|Ice Storm, The (1...|      0|         (     |Heavenly Creature...|      0|         0|         0|
|     305|Ice Storm, The (1...|      0|         (
```

> Did you realize by now that on a big data dataset this operation is massive and requires extensive computing resources. Thankfully on Spark you can distribute this operation to run on multiple computer notes. For this, you can tweak the `spark-submit` job with the following parameters so as to extract the last bit of juice from all the clustered nodes:
>
> ```
> spark-submit executor-cores <Number of Cores>
> --num-executor <Number of Executors>
> ```

The next step is the most important one in this content management program. We will run over the results of the previous query and from each row we will pull out the data for movie1 and movie2 and cross compare the attributes of both the movies and find the Euclidean Distance between them. This would show how similar both the movies are; the greater the distance, the less similar the movies are. As expected, we convert our dataset to an RDD and invoke a map function and within the `lambda` function for that map we go over all the dataset rows and from each row we extract the data and find the Euclidean Distance. For maintaining conciseness, we depict only a portion of the following code. Also note that we pull the information for both movies and store the calculated Euclidean Distance in a `EuclidVO`Java POJO object:

```
JavaRDD<EuclidVO> euclidRdd = movieDataDS.javaRDD().map( row -> {
    EuclidVO evo = new EuclidVO();
        evo.setMovieId1(row.getString(0));
        evo.setMovieTitle1(row.getString(1));
        evo.setMovieTitle2(row.getString(22));
        evo.setMovieId2(row.getString(21));
```

```
            int action = Math.abs(Integer.parseInt(row.getString(2)) -
Integer.parseInt(row.getString(23)) );
    ...
double likesCnt = Math.abs(row.getDouble(20) - row.getDouble(41));

        double euclid = Math.sqrt(action * action + ... + likesCnt *
likesCnt);
            evo.setEuclidDist(euclid);
        return evo;
});
```

As you can see in the bold text in the preceding code, we are calculating the Euclid Distance and storing it in a variable. Next, we convert our RDD to a dataset object and again register it in a temporary view `movieEuclids` and now are ready to fire queries for our predictions:

```
Dataset<Row> results = spark.createDataFrame(euclidRdd.rdd(),
EuclidVO.class);
results.createOrReplaceTempView("movieEuclids");
```

Finally, we are ready to make our predictions using this dataset. Let's see our first prediction; let's find the top 10 movies that are closer to Toy Story, and this movie has `movieId` of 1. We will fire a simple query on the view `movieEuclids` to find this:

```
spark.sql("select * from movieEuclids where movieId1 = 1 order by
euclidDist
asc").show(20);
```

As you can see in the preceding query, we order by Euclidean Distance in ascending order as lesser distance means more similarity. This would print the first 20 rows of the result as follows:

```
+-------------------+--------+--------+----------------+--------------------+
|       euclidDist|movieId1|movieId2|      movieTitle1|         movieTitle2|
+-------------------+--------+--------+----------------+--------------------+
|1.0021449993740907|       1|      95|Toy Story (1995)|      Aladdin (1992)|
| 1.003062211735072|       1|     969|Toy Story (1995)|Winnie the Pooh a...|
|1.0208065527493175|       1|     404|Toy Story (1995)|    Pinocchio (1940)|
|1.0256053961344038|       1|     189|Toy Story (1995)|Grand Day Out, A ...|
|  1.03216473791695|       1|     422|Toy Story (1995)|Aladdin and the K...|
| 1.140813842164051|       1|     625|Toy Story (1995)|Sword in the Ston...|
| 1.159952147889726|       1|     169|Toy Story (1995)|Wrong Trousers, T...|
|1.2024751613136604|       1|     477|Toy Story (1995)|     Matilda (1996)|
|1.2385620296700004|       1|     946|Toy Story (1995)|Fox and the Hound...|
+-------------------+--------+--------+----------------+--------------------+
```

As you can see in the preceding results, they are not bad for content recommender system with little to no historical data. As you can see the first few movies returned are also famous animation movies like Toy Story and they are Aladdin, Winnie the Pooh, Pinocchio, and so on. So our little content management system we saw earlier is relatively okay. You can do many more things to make it even better by trying out more properties to compare with and using a different similarity coefficient such as Pearson Coefficient, Jaccard Distance, and so on.

> If you run this program on your computer or laptop from our GitHub repository, you would realize that it is not a very fast program. This is because it involves too many comparisons between the datapoints to find the similarity between them. As such we need a more efficient algorithm to calculate the recommendations. Next we will see one such algorithm for making faster recommendations.

Collaborative recommendation systems

As the name suggests, these kinds of recommendation systems are based on collaborative inputs from different users, their browsing criteria, the ratings they gave to different products, the products they purchased, and so on. The idea is that if we have historical data of the users' browsing history, then based on that browsing history we can figure out patterns that depict the choice or preference of the users and based on that we can find other people with similar preferences. Now, we can propose new products to these people (that they have not checked yet) based on the fact that the users who are similar to them have shown preference for them. Let's try to understand the concept of collaborative filtering (this is the name of the approach) using the following diagram:

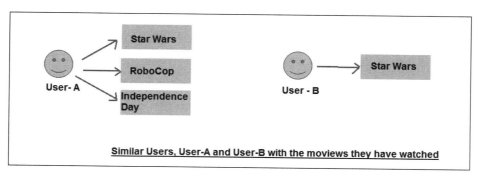

Similar Users, User-A and User-B with the moviews they have watched

As you can see in the preceding diagram the **User-A** watched and liked three movies but **User-B** which is similar to **User-A** (as **User-B** rated the movie **Star Wars** as highly as **User-A** did) has watched only one. So the two movies from **User-A** that **User-B** has not watched can be recommended to **User-B**. Thus, a collaborative system would propose User-B to watch **Independence Day** and **RoboCop**. This is a very simplistic depiction of collaborative filtering, but the base idea depicted is correct.

This is all what collaborative filtering is all about. In short, it can be depicted in the following few steps:

- A user views, browses, rates, or buys on a website. All these transactions of the user are recorded

- From these transactions, patterns are then extracted to figure out the preferences of the users

- Next, the users with similar preferences are figured out by cross matching their preferences to each other and finding the similarity between them (recall the mathematical formulas for similarity shown in the preceding content-based systems)

- Similar users thus figured out are now recommended with products that they have not browsed, or rated yet based on the fact that other similar users have used these products. Since they are similar to each other, they can be shown each others' choices (products), which they have not seen yet, as it is assumed that being of similar taste they might like that too

> In the preceding example we covered an example of user-based collaborative filtering. However, there is another approach of Item Based Collaborative Filtering. It does comparison between item to item based on collaborative data, that is, past transactions data based on the items or products.

There are some clear advantages and disadvantages of collaborative filtering over content-based recommenders and they are listed next.

Advantages

- They have clear advantage of showing recommendations to users that might not be themselves related. For example, if a user likes star wars movie and also likes playing the video game Halo, the user might be presented with this recommendation on the e-commerce sites. But the same is not true with content managed systems, as the two products are not related at all—one is a movie and the other is a video game.

- Since they don't require product comparison among each other as they are based on other users past data (like ratings they gave), they can be made to be very fast specially on big data when cluster computing frameworks such as Apache Spark is used.

Disadvantages

- They suffer from the clear disadvantage of cold start problem. Basically, they require some user data initially to work on as such on totally new websites with no user history of data, they won't be useful.
- They do not take the actual content of the items into account when making recommendation and hence are not content aware.

A recommender system directly impacts the business and as such it is used in almost all big web stores such as Amazon, Netflix, and so on. As such the main use case or real-world usage of these systems come in places where plenty of data is involved, in other words big data is involved. We have now seen the basic concept of collaborative filtering, we will now try to understand how it is used on big data specially using Apache Spark cluster computing.

Alternating least square – collaborative filtering

Apache Spark provides an inbuilt algorithm for collaborative filtering and it is based on the approach of alternating least square. Before we look deeply into this approach, let's look at the basic concept of matrix factorization. This is an old algebraic approach for breaking or decomposing a big matrix into smaller individual matrices. These smaller matrices can be multiplied together to produce the bigger matrix. Let's try to understand this using a simple example shown next.

On the left-hand side, we have a big matrix and on the right-hand side we have two smaller matrices that when multiplied together produce the bigger matrix. This is what is matrix factorization. We have split the bigger matrix into two smaller matrices for easier and better representation:

$$\begin{bmatrix} 1 & 2 & 1 \\ 2 & 4 & 2 \\ 3 & 4 & 3 \\ 2 & 4 & 2 \end{bmatrix} = \begin{bmatrix} 1 \\ 2 \\ 3 \\ 2 \end{bmatrix} * \begin{bmatrix} 1 & 2 & 3 \end{bmatrix}$$

As it is an inbuilt algorithm in Apache Spark, the code for this is straightforward, yet it is very good in terms of predictive results and also this algorithm is highly scalable, so various portions of the algorithm can be parallelly run on multiple nodes.

The concept of the algorithm is simple and is shown by the following steps:

1. **Get the original matrix of the user and item combinations**: Thus, on the rows you will have users and on the columns you will have the items (for example, movies). The entry into the row and column of the matrix would depict the rating or any other entity the user has assigned to the item (or movie). An example of such a matrix is shown as follows:

	Movie-1	Movie-2	Movie-3	Movie-4	Movie-5
User-1	1	5	?	?	3
User-2	?	?	?	2	5
User-3	2	?	3	2	?
User-4	?	4	3	4	3
User-5	5	2	?	?	?

As you can see in the preceding matrix, the rows are represented for the users and the columns are represented for the movies. The entries within the matrix are the values of *rating* given to the movies by the users. The ? sign depicts missing entries where the user has not rated the movies. So for predictive analytics algorithm, the job is to predict the rating of movies for these missing entries and the highest rated movies for the prediction will become the recommendation for the users.

From the perspective of collaborative filtering, let's understand the preceding matrix. Let's look at the two users **User-1** and **User-4**. Both these users have rated the movie **Movie-5** with rating **3** and **Movie-2** with a high rating of **5** and **4** respectively. Thus it does look like both these users have a similar taste, as they have near similar ratings for movies they watched. Now **User-4** has watched two extra movies that is, **Movie-3** and **Movie-4** which **User-1** has not watched, so these movies can be recommended to **User-1**. Also to find the similarity between **User-1** and **User-4** and also between other users, Pearson Coefficient, Euclidean Distance, and so on can be used.

Recommendation Systems

2. **Do matrix factorization on this sparse matrix and break it into two smaller matrixes**: This matrix is sparse and huge and on a massive big data dataset level it will have many missing entries. It's a good idea to use matrix factorization and convert it into dense smaller matrixes. The matrix can be broken into smaller matrixes as shown next:

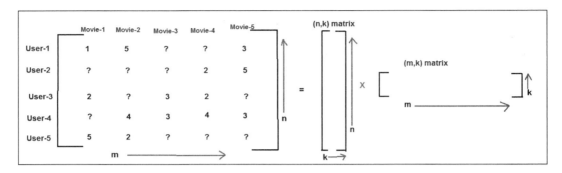

As you can see in the preceding diagram, a large matrix is broken (factorized) into smaller matrixes. The first $n \times k$ matrix is for the users and the next $m \times k$ matrix is for the movies. These matrixes are filled with the numbers that we figured out and the multiplication of these matrixes would give the matrix on the left-hand side.

3. **To figure out the rating for a user at index u of the matrix and for a movie at index i**: This is easy, just grab the value at the u location from the (n,k) matrix and value at the i location from the (m,k) matrix and multiple them together. This would give us the predicted rating. This way we can figure out the predicted rating for all the missing entries.

We saw how this approach can be used to figure out the predictive results. One important piece we have not covered yet is, "How do we figure out the factorized matrixes?"

We would try to explain the concept of factorization in simple words. As you recall that we initially broke our large user/movies matrix into two smaller manageable user factors (we call it user factors) matrix and movie factors matrix. Next we need to figure out the values in these user factors and movie factors matrix.

An approach for this is to assign a high value; (since our max rating is 5) we can assign and initialize the user factor matrix with these values and as for k, we can randomly choose some value for k say 10. Next, we keep this x factors matrix as constant and we now figure out the values in the movie factors matrix. Next, we multiply these matrixes and figure out the mean squared error. Now, our task is to figure out the best fit values in the user factor matrix and movie factors matrix so that the mean squared error is minimized. To avoid overfitting, the concept of regularization is also introduced in the formula (so a certain level of bias is added so as to avoid overfitting, just consider this as an optimization technique so that our model does not get overfitted). Now this was the first step, next what we do is we keep the calculated movie factors matrix as a constant and calculate the user factors matrix again using the minimizing function of mean squared error and regularization combination. We alternate (hence the name alternating) between the matrixes by keeping one constant and figuring out the values of the other one at a time while constantly minimizing the error value. Finally, we will come at the values of user factors and movie factors beyond which the error difference (mean squared error) won't change much and hence these will be the best figured out user factors and movie factors matrixes.

From the perspective of Spark, you might have figured out that these alternate matrix computations can be done parallelly on multiple nodes and hence it is chosen as part of the default algorithm in Spark for recommendations. Also, this algorithm gives very good results even in case of very sparse user/movie matrixes.

We have omitted the math involved in the calculation of this ALS algorithm simply to keep the complexity low in the preceding text as we believe this to bean introductory book for Java developers in analytics. For more details on this topic, we would urge the users to check out the research papers on the ALS algorithm in sites such as Google scholar or `arvix.org`.

Here comes the main piece of this algorithm.

Let's start going over the code of the algorithm now.

First is the usual boiler plate code to create the `SparkSession` object. For this we first build the spark configuration object and provide the master which in our case is 'local' since we are running this locally in our computer. Next using this spark configuration object we build the `SparkSession` object and also provide the name of the application here and that is `ContentBasedRecommender`.

```
SparkConf sc = new SparkConf().setMaster("local");
SparkSession spark = SparkSession
            .builder()
```

Recommendation Systems

```
            .config(sconf)
            .appName("CollaborativeRecommendMovies")
            .getOrCreate();
```

Using this `SparkSession` object, now fetch the data from the file containing the movies data (their ID and their respective titles). We will pull this data in a Java RDD object and fire a `map` function on that RDD and pull the data into a Java value object called `MovieVO`:

```
JavaRDD<MovieVO> movieRdd = spark
                        .read().textFile("data/movie/u.item").
javaRDD()
                        .map(row -> {
                        String[] strs = row.split("\\|");
                            MovieVO mvo = new MovieVO();
                        mvo.setMovieId(strs[0]);
                        mvo.setMovieTitle(strs[1]);
                return mvo;
                        });
```

We will convert this RDD into a dataset and also register it as a temporary view as we will be later needing it while firing predictive queries:

```
Dataset<Row> movieDS = spark.createDataFrame(movieRdd.rdd(), MovieVO.
class);
        movieDS.createOrReplaceTempView("movies");
```

We will now fetch the data from the file containing the user movie ratings values (`u.data` file in dataset). After the file is loaded into the RDD, run a `map` function on the RDD and using this function from each row extract the data for the user ratings and store in a `RatingVO` Java value object:

```
JavaRDD<RatingVO> ratingsRDD = spark
    .read().textFile("data/movie/u.data").javaRDD()
    .map(row -> {
    String[] fields = row.split("\t");
        if (fields.length != 4) {
            return null;
        }
        int userId = Integer.parseInt(fields[0]);
        int movieId = Integer.parseInt(fields[1]);
        float rating = Float.parseFloat(fields[2]);
    long timestamp = Long.parseLong(fields[3]);
            return new RatingVO(userId, movieId, rating, timestamp);
}).filter(f -> f != null);
```

As you can see, whenever a missing data field is obtained (in the lambda function where we check for lesser parameters), we return a null value and we filter out this row. Thus, for this use case, we are filtering and removing missing data.

Next, convert this Spark RDD containing the Java rating POJO objects into a dataset using the `createDataFrame` method and using the `randomsplit` method, split this dataset into training dataset and test datasets separately:

```
Dataset<Row> ratings = spark.createDataFrame(ratingsRDD, RatingVO.class);
Dataset<Row>[] splits = ratings.randomSplit(new double[]{0.8, 0.2});
Dataset<Row> training = splits[0];
Dataset<Row> test = splits[1];
```

Now it's time to build the recommendation model. We will use the Apache Spark that provides **Alternating Least Square** (**ALS**) algorithm from the Spark ML library. This algorithm, as we discussed in a previous section, is an example of collaborative filtering. After the instance of the algorithm ALS is created, we supply the necessary parameters to it:

```
ALS als = new ALS()
        .setMaxIter(10)
        .setRegParam(0.01)
        .setUserCol("userId")
        .setItemCol("movieId")
        .setRatingCol("rating");
```

The specific parameters are explained in the following table:

Maximum Iterations	This is the maximum number of iterations to run; the default value is 10
Reg Param	This specifies the regularization parameter; the default value is 1.0
User Column	This is the column to read the user ID from
Item Column	This is the column to read the item ID or movie ID from
Rating Column	This is the column to read the explicit rating from, given by the user

Once our model instance is built and configured, next is the time to fit the model on the training data. This would help the model internally and build its knowledge base, based on which it can later make predictions:

```
ALSModel model = als.fit(training);
```

Recommendation Systems

Now it's time to run the model on test data, using the transform function on the model. We will also run the show method to print the first few lines of the predicted results:

```
Dataset<Row> predictions = model.transform(test);
predictions.show();
```

After running the show method, the results would be printed as follows:

```
+-------+------+---------+------+---------+
|movieId|rating|timestamp|userId|prediction|
+-------+------+---------+------+---------+
|    148|   1.0|875326138|   633| 3.8584247|
|    148|   3.0|886106165|   271| 2.7496443|
|    148|   3.0|878150506|   606| 3.4842434|
|    148|   4.0|890117028|   236|  3.028388|
|    148|   3.0|876348140|   601| 1.1929333|   <=== Predictions for User Ratings
|    148|   3.0|888104154|   224|  4.195447|
|    148|   2.0|887160606|   896| 2.6896698|
|    148|   4.0|882824325|   178| 3.8263626|
|    148|   3.0|887740788|   308| 2.5620613|
|    148|   4.0|880387474|   923| 3.4326193|
|    148|   3.0|889490499|   120| 2.3387685|
|    148|   2.0|877226047|   430|  2.688941|
|    148|   2.0|877383934|    92| 2.5610256|
|    148|   4.0|878854729|   447|  3.031723|
```

As you can see in the preceding algorithm, it is making predictions about what would the user rate the movies as. It's easy to figure out that whatever predicted ratings are high in value, the user might like these movies. Let's now find out one random user, say userId equal to 633 and for this user let's find the movies for which predicted ratings are greater than 3, and we would assume that the user would like these movies.

For this, we would register the preceding predictions dataset as a temporary view called predictions and fire a simple Spark SQL select query on it as shown next. In the query, we would sort the results by showing the highest rated predictions first:

```
predictions.createOrReplaceTempView("predictions");
spark.sql("select m.movieTitle,p.* from predictions p,movies m where p.userId =
633 and p.movieId = m.movieId order by p.prediction desc").show();
```

As you can see in the preceding query, we also needed a join with the movie table so that we could extract the actual movie name from it.

This would print the predicted results for the user as follows:

```
+--------------------+----+-------+------+---------+------+---------+
|          movieTitle|like|movieId|rating|timestamp|userId|prediction|
+--------------------+----+-------+------+---------+------+---------+
|Murder in the Fir...|   0|    939|   4.0|877212045|   633| 4.1661263|
|Terminator 2: Jud...|   0|     96|   4.0|875324997|   633|   4.10481|
|    Home Alone (1990)|  0|     94|   4.0|877211684|   633| 3.7664657|
|Fugitive, The (1993)|   0|     79|   5.0|875325128|   633| 3.6715062|
|Schindler's List ...|   0|    318|   4.0|875324813|   633| 3.6350746|
|        Ben-Hur (1959)|  0|    526|   4.0|877212250|   633| 3.5871043|
|           Jaws (1975)|  0|    234|   4.0|877212594|   633| 3.5124974|
|Good, The Bad and...|   0|    177|   3.0|875325654|   633| 3.3321676|
|         Scream (1996)|  0|    288|   2.0|875324233|   633| 3.3247242|
|       Supercop (1992)|  0|    128|   3.0|875325225|   633| 3.2833974|
|          Glory (1989)|  0|    651|   3.0|877212283|   633| 2.9536574|
```

Predicted Ratings

Predicted Movies

We have seen the predicted results, let's now see how good our predictions are. For this, we will use the `RegressionEvaluator` class provided by Apache Spark. We will be concentrating on the parameter `rmse` or root mean squared error (which is nothing but mainly the difference between the predicted and actual value). We will provide this metric to the regression evaluator, that is, `rmse` and the actual and predicted columns. We would next run the evaluations on the predicted or test dataset and print the value of error to console:

```
RegressionEvaluator evaluator = new RegressionEvaluator()
                                .setMetricName("rmse")
                                .setLabelCol("rating")
                                .setPredictionCol("prediction");
Double rmse = evaluator.evaluate(predictions);
       System.out.println("Root-mean-square error = " + rmse);
```

Try to tweak the preceding parameters to get better results in terms of accuracy from the recommendation model.

The example we just depicted for the recommendations on top of the MovieLens dataset is an example of explicit dataset. In this dataset, we had explicit entries of ratings by the users and using that we were easily able to gauge at the predictions. However, if you check in real life how many times did you rate a YouTube video after watching it. It's a very rare thing to do and many users don't rate the videos after watching it. Hence, we need more and different parameters for making predictions. These parameters can be something like the number of times the movie is watched, the number of people who add it to their wish list, the number of people who ordered the DVD, and so on. The model from the Spark ALS model supports the notion of implicit dataset and we can specify that as a parameter and use the ALS model from Spark for implicit datasets too.

With this, we come to an end to our chapter on recommendation engines. We urge the users to read more documentation on this on the web or on research papers such as `arvix.org`—this is a very upcoming and growing area of work.

Summary

In this chapter, we learned about recommendation engines. We saw the two types of recommendation engines, that is, content recommenders and collaborative filtering recommenders. We learned how content recommenders can be built on zero to no historical data and are based on the attributes present on the item itself, using which, we figure out the similarity with other items and recommend them. Later, we worked on a collaborative filtering example using the same MovieLens dataset and the Apache Spark alternating least square recommender. We learned that collaborative filtering is based on historical data of users' activity, based on which other similar users are figured out and the products they liked are recommended to the other users.

In the next chapter, we will learn two important algorithms that are part of the unsupervised learning world and they will help us form clusters or groups in unlabeled data. We will also see how these algorithms help us segment the important customers in an e-commerce store.

10
Clustering and Customer Segmentation on Big Data

Up until now we have only used and worked on data that was prelabeled that is, supervised. Based on that prelabeled data, we trained our machine learning models and predicted our results. But what if the data is not labeled at all and we just get plain data? In that case, can we carry out any useful analysis of the data at all? Figuring out details from an unlabeled dataset is an example of unsupervised learning, where the machine learning algorithm makes deductions or predictions from raw unlabeled data. One of the most popular approaches to analyzing this unlabeled data is to find groups of similar items within a dataset. This grouping of data has several advantages and use cases, as we will see in this chapter.

In this chapter, we will cover the following topics:

- The concepts of clustering and types of clustering, including k-means and bisecting k-means clustering
- Advantages and use cases of clustering
- Customer segmentation and the use of clustering in customer segmentation
- Exploring the UCI retail dataset
- Sample case study of clustering for customer segmentation on the UCI retail dataset

Clustering

A customer using an online e-commerce store to buy a phone would generally type those words in the search box at the top of the site. As soon as you type your search query, the search results are displayed at the bottom, and on the left-hand side of the page you get a list of categories that you might be interested in based on the search text you just entered. The sub-search categories are shown in the following screenshot. How did the search engine figure out these sub-search categories just based on the searched text? Well, this is what **clustering** is used for. It's a no-brainer that the site's search engine is advanced and must be using some form of clustering technique to group the search results so as to form useful sub-search categories:

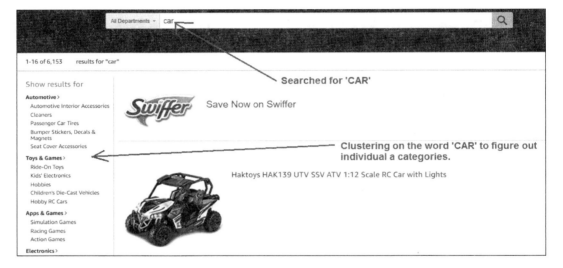

As seen in the preceding screenshot, the left-hand side shows the categories (groups) that are generated once the user searches for a term such as car. The left-hand side looks quite relevant as we are seeing sub-categories for car accessories, **Toys & Games**, **Electronics**, and so on.

Let's try to define clustering in formal terms. Clustering is a popular form of an unsupervised learning algorithm and is used to discover and club similar entities (like movies, customers, products, likes, and so on) into groups or clusters. Within a group, the items are similar to each other based on certain attributes they possess. To depict clustering visually, let's look at the scatter plot shown as follows. This scatter plot shows the prices of houses versus their living area:

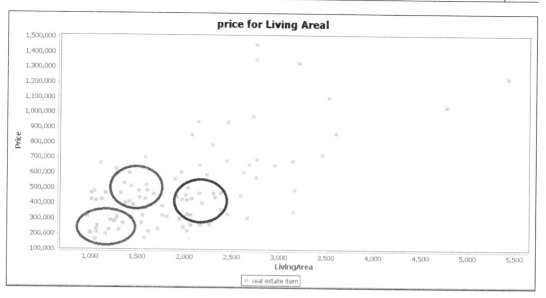

As you can see in the preceding graph, there are three ellipsoids. These ellipsoids depict clustered groups within these datasets. Within these clusters, the data points selected will be similar. The similarity of the datasets is calculated based on the similarity finding mathematical formulas and approaches that we discussed earlier, like Euclidean distance, Pearson coefficient, Jaccard distance, cosine similarity, and so on.

Though we have shown just one example previously, clustering in general has many practical uses and is used in many real-world use cases, as follows:

- **Customer segmentation**: Clustering can be used to segment customers, for example, of online e-commerce stores.
- **Search engines**: Clustering is used to form smaller groups of sub-search categories that can help users looking for specific items via a search engine. We covered these use cases in the introduction of clustering.
- **Data exploration**: In the data exploration phase, clustering can be used.
- **Finding epidemic breakout zones**: There are studies where clustering has been applied to find zones of disease breakout areas based on the availability of statistical data.
- **Biology**: To find groups of genes that show similar behavior.
- **News categorization**: Google News does this to categorize news automatically into groups such as sports news, technology related news, and so on.

- **Summarization of news**: First, we can find custom groups in the news and then find the centroid and use it to summarize the news.

Even though we have listed a few sample use cases of clustering, we have just scratched the surface of clustering. There are plenty of other uses of clustering in real life.

Now that we know what clustering is, let's see the type of clustering we can perform on big data.

Types of clustering

There are many types of clustering algorithms, but for the purpose of this chapter we will only be dealing with the three main types of clustering, and they are as follows:

- Hierarchical clustering
- k-means clustering
- Bisecting k-means clustering

Next, we will cover each one of these in their own sub-section.

Hierarchical clustering

This is the simplest form of clustering and can be explained using the following steps:

1. The concept of hierarchical clustering is very simple; first, we find the distance between all the data points based on their attributes. To find the distance we can use any algorithm like the Euclidean distance.

2. Next, we find the two most similar points based on this distance and combine them in a group.

3. Next, we find the average or centroid of this group and, using this centroid we again find the distance of this point with respect to other points in the dataset. The closest point is again found and the previous cluster is now combined with the new data point to form a new cluster.

4. This process keeps on repeating until we find the required number of clusters we are looking for.

Hierarchical clustering is explained using the following diagram:

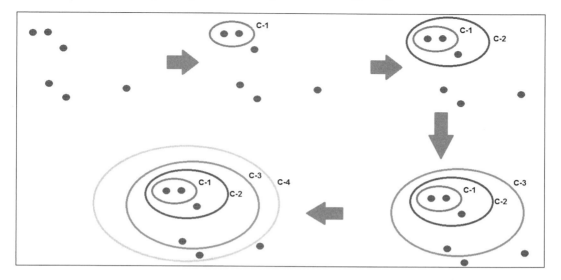

As can be seen in the preceding diagram, in the first image we just have a few data points, then, as mentioned in the hierarchical clustering steps previously, we figure out the distance between all the points. Between the shortest distance between two points we build our first cluster, and this is shown within the smallest ellipsoid in the second image (that is, cluster **C-1**). Next, we find the centroid of this cluster, and using that centroid, we again find the distance with respect to all the other points find the shortest distance from this centroid and draw a circle or mark the cluster. This time we build the cluster **C-2**, as shown in the third image from the left above. This process continues until we figure out the other clusters **C-3** and **C-4**.

Hierarchical clustering gives very good results in terms of cluster accuracy (that is, clubbing similar items within a cluster), however as you may have guessed, hierarchical clustering is a highly computationally intensive approach, as you have to perform a lot of calculations to figure out the similarity between the data points. On a big data dataset, this is quite infeasible as too many computations would slow down this approach tremendously. On a small dataset, this approach can be used with good results. As this approach doesn't make much sense with a big data dataset, we won't be covering this approach further. Next, we will look into one of the most popular clustering forms known as k-means clustering.

K-means clustering

As we saw in the previous diagrams, on a scatter plot, the data points that were similar in nature were usually placed close to each other on that chart. This is the main idea behind k-means clustering, where *k* random centroids of clusters are picked, and based on those centroids their nearest data points are figured out and put in a group or cluster. Thus, the name k-means clustering. To optimize this clustering process further, centroids of clusters that have been discovered are calculated by taking the average of all the other data point attributes. These k-means clustering average points or new centroids are then used as the new *k* points, and the nearest points are figured out to form new *k* clusters. This process is repeated until you cannot change the value of these centroids (the average values) further.

The whole concept of k-means clustering is depicted in the following diagram:

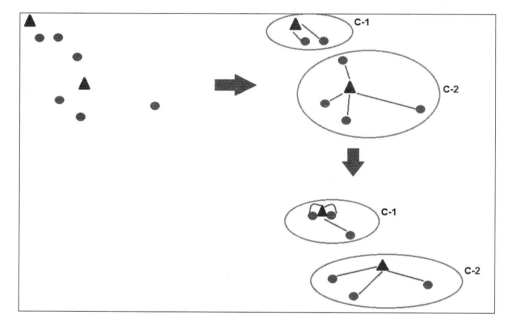

As we saw in the previous diagram, our task here is to find just 2 clusters (so *k* = 2) in these data points. The forming of clusters in the previous diagram is done with the following steps:

1. First, the data points of the dataset are plotted on a scatter plot (this is covered by the leftmost image with the dots depicting the data points).
2. Next, the two (that is, *k*) cluster centroids are chosen randomly. These are shown by the triangles.

3. After choosing the cluster centroids randomly, the nearest datapoints are figured out and put into two cluster groups. These clusters are named **C-1** and **C-2** in the previous diagram and are also shown using ellipsoids. The ellipsoids cover the datapoints within them.

4. To optimize the clusters further, in the last diagram on the bottom right-hand side, the triangles are put on the average points of the previous clusters, the distances are now figured out from these points to the other points, and the clusters are redrawn. You should now be able to see that the C-1 cluster had just two data points, but now has three.

> To calculate the distance between the data points, any mathematical approach like Pearson coefficient, Euclidean distance, and so on, can be used.

K-means clustering is a much faster approach than hierarchical clustering as it does not require the program to calculate the similarity coefficients between data points again and again. Due to this, it can be used to run on very large datasets. This is the main reason k-means clustering is built into the machine learning library of Spark and is a standard algorithm that ships with it.

There is one drawback of k-means clustering and that is the randomness of choosing the *k* clusters. So, if next time you randomly choose some other k points, you would get a different set of clusters.

In the next section, we will see another form of clustering that can give better performance than k-means clustering.

Bisecting k-means clustering

This clustering technique is a combination of hierarchical clustering and k-means clustering. The main purpose of this clustering approach is to enhance the accuracy bottleneck of the k-means clustering. This algorithm is more accurate than the normal k-means clustering algorithm and is faster than the hierarchical clustering algorithm as it requires fewer iterations.

The concept of this technique is simple: it first starts by putting all the datapoints into a single cluster. Next, within that cluster, it further bifurcates the data into two smaller clusters using k-means clustering, that is, by taking *k* = 2. From these two clusters, it figures out the **sum of mean squared errors** (**SMEs**). It picks the cluster with the lowest mean squared errors, that is, it picks the cluster that has the maximum similarity between the items. The algorithm is made to run for a fixed number of times until the required numbers of clusters is reached.

The technique is depicted in the following diagram:

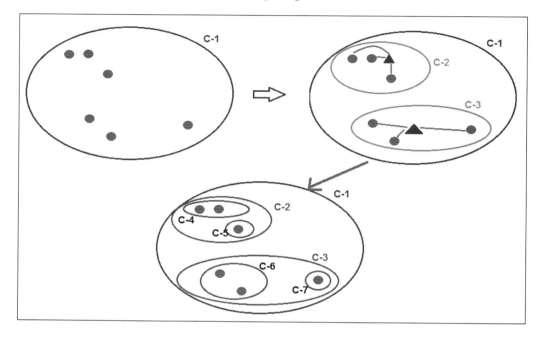

As seen in the previous diagram, the bisecting k-means happened on the sample dataset (shown in the scatter plot with data points on the left-hand side on top). The bisecting k-means steps that occurred in the previous diagram can be explained as follows:

1. First, all the data points were put in a single cluster called **C-1**.
2. Next, the **mean squared error** (**MSE**) of all the datapoints within this cluster were calculated.
3. After this, **C-1** was bifurcated into two clusters, **C-2** and **C-3** (as shown in the second diagram on the right-hand side). K-means clustering was used to bifurcate **C-1** into two sub-clusters, **C-2** and **C-3**.

> To choose the sub-clusters C-2 and C-3, we calculate the total MSE of the two sub-clusters formed. The clusters with the lowest mean squared error are the most similar and, therefore, are selected.

4. We repeat the step of bifurcation on the sub-clusters until the total number of expected clusters is reached.

As you saw previously, this approach is a combination of hierarchical and k-means. It gives comparable results to hierarchical clustering and is much better than the pure randomness involved in k-means clustering.

We have now seen some of the most common clustering techniques, so let's now utilize them on a case study for customer segmentation.

Customer segmentation

Customers for any store either offline or online (that is, e-commerce) all exhibit different behaviors in terms of buying patterns. Some might buy in bulk, while others might buy lesser quantities of stuff but the transactions might be spread out throughout the year. Some might buy big items during festival times like Christmas and so on. Figuring out the buying patterns of the customers and grouping or segmenting the customers based on their buying patterns is of the utmost importance for the business owners, simply because it lays out the customers' needs in front of them and their importance. They could selectively market to the more important customers, thereby giving prime care and importance to the customers that generate maximum revenue for the stores.

Figuring out the buying patterns of the customers from historical data (of their purchase transactions) is easy for an online store as all the transaction data is readily available. Some approaches that people use for this are easy like firing some simple queries on their databases to generate reports, but in this chapter, we will be discussing an advanced approach to customer segmentation that has become popular in the recent years. **Recency**, **Frequency**, and **Monetary** (**RFM**) analysis is an approach we are going to use to figure out our important customers and their buying patterns, and later we will use this to segment or group them using clustering algorithms.

RFM and the terms are explained in the following table:

Recency	This value depicts how recently the customer has purchased. We will be using this as the most recent purchase date and calculating the difference between the current date and this date, so the net value for this will be an integer representing the number of days since last purchase.
Frequency	This value depicts how many times the customer has purchased. We will be using it as the count of the number of times the customer has purchased within a year.
Monetary	This value depicts the amount the customer has spent overall in the store. We will store this as the amount depicted by multiplying the cost of items and the quantity sold to the customer.

Dataset

For our case study on customer segmentation using clustering, we will be using a dataset from UCI repository of datasets for a UK online retail store. This retail store has shared its data with UCI and the dataset is freely available on their website. This data is essentially the transactions of different customers made on the online retail store. The transactions were made from different countries and the dataset size is good (thousands of rows). Let's go through the attributes of the dataset:

Attribute name	Description
Invoice number	Invoice number; a number uniquely assigned to each transaction
Stock code	Product (item) code; a 5-digit integral number uniquely assigned to each distinct product
Description	Product item name
Quantity	Quantity of items purchased in a single transaction
Invoice date	Date of the transaction
Unit price	Price of the item (in pounds)
Customer ID	Unique ID of the person making the transaction
Country	Country from where the transaction was made, such as United Kingdom or France

Before we dive into customer segmentation for this dataset, let's explore this data first and learn about it.

Data exploration

In this section, we will explore this dataset and try to perform some simple and useful analytics on top of this dataset.

First, we will create the boilerplate code for Spark configuration and the Spark session:

```
SparkConf conf = ...
SparkSession session = ...
```

Next, we will load the dataset and find the number of rows in it:

```
Dataset<Row> rawData = session.read().csv("data/retail/Online_Retail.csv");
```

This will print the number of rows in the dataset as:

```
Number of rows --> 541909
```

As you can see, this is not a very small dataset but it is not big data either. Big data can run into terabytes. We have seen the number of rows, so let's look at the first few rows now.

```
rawData.show();
```

This will print the result as:

```
17/05/16 08:50:06 INFO CodeGenerator: Code generated in 32.169375 ms
+------+------+--------------------+---+--------------+----+-----+--------------+
|   _c0|   _c1|                 _c2|_c3|           _c4| _c5|  _c6|           _c7|
+------+------+--------------------+---+--------------+----+-----+--------------+
|536365|85123A|WHITE HANGING HEA...|  6|12/1/2010 8:26|2.55|17850|United Kingdom|
|536365| 71053| WHITE METAL LANTERN|  6|12/1/2010 8:26|3.39|17850|United Kingdom|
|536365|84406B|CREAM CUPID HEART...|  8|12/1/2010 8:26|2.75|17850|United Kingdom|
|536365|84029G|KNITTED UNION FLA...|  6|12/1/2010 8:26|3.39|17850|United Kingdom|
|536365|84029E|RED WOOLLY HOTTIE...|  6|12/1/2010 8:26|3.39|17850|United Kingdom|
|536365| 22752|SET 7 BABUSHKA NE...|  2|12/1/2010 8:26|7.65|17850|United Kingdom|
|536365| 21730|GLASS STAR FROSTE...|  6|12/1/2010 8:26|4.25|17850|United Kingdom|
|536366| 22633|HAND WARMER UNION...|  6|12/1/2010 8:28|1.85|17850|United Kingdom|
|536366| 22632|HAND WARMER RED P...|  6|12/1/2010 8:28|1.85|17850|United Kingdom|
|536367| 84879|ASSORTED COLOUR B...| 32|12/1/2010 8:34|1.69|13047|United Kingdom|
|536367| 22745|POPPY'S PLAYHOUSE...|  6|12/1/2010 8:34| 2.1|13047|United Kingdom|
|536367| 22748|POPPY'S PLAYHOUSE...|  6|12/1/2010 8:34| 2.1|13047|United Kingdom|
|536367| 22749|FELTCRAFT PRINCES...|  8|12/1/2010 8:34|3.75|13047|United Kingdom|
```

As you can see, this dataset is a list of transactions including the country name from where the transaction was made. But if you look at the columns of the tables, Spark has given a default name to the dataset columns. In order to provide a schema and better structure to the data, we will convert and store the data per row into a Java object (JavaBean). We will cover this code in detail in the next section for actual clustering. But for now, we will just show you that we have converted and stored the data into a Java RDD of **Plain Old Java Object** (**POJO**) and converted it into a dataset object:

```
JavaRDD<RetailVO> retailData = …
Dataset<Row> retailDS = session.createDataFrame(retailData.
rdd(),RetailVO.class);
```

We will register this dataset as a temporary view so that we can fire data exploration queries on it:

```
retailDS.createOrReplaceTempView("retail");
```

Next, we fire our first query to group the number of data rows by country in this dataset. We are not interested in data rows where the count is less than one thousand (as this dataset is large, this is a good case):

```
Dataset<Row> dataByCtryCnt = session.sql("select country,count(*) cnt
from retail group by country having cnt > 1000");
        dataByCtryCnt.show()
```

By invoking a show method on this dataset, as shown previously, the following results will be printed:

```
+-----------+------+
|    country|   cnt|
+-----------+------+
|    Germany|  9495|
|     France|  8491|
|    Belgium|  2069|
|       EIRE|  7485|
|     Norway|  1086|
|      Spain|  2533|
|         UK|361878|
|Switzerland|  1877|
|   Portugal|  1480|
|  Australia|  1259|
|Netherlands|  2371|
+-----------+------+
```

The data we got in the previous output is best explained through a bar chart. We will show the chart as follows. We will not show you the code for the chart, to maintain brevity, but it is part of our GitHub repository:

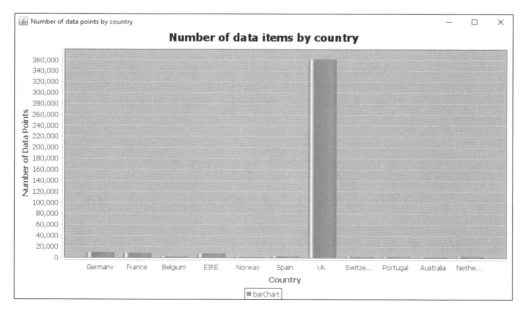

As you can see, the previous transaction data in this dataset contains the maximum number of data items from the United Kingdom. Thus, it makes sense to concentrate and perform clustering only on this data for our simple case study. We will filter the country to the United Kingdom, so let's now discover the basic stats about the sold items, for example, the average price or maximum price of items sold, as well as the average or maximum quantity sold.

To do this, we will fire a simple select query on all the columns, filter on United Kingdom, and finally run a describe and show on the dataset obtained, as follows:

```
Dataset<Row> ukTransactions = session.sql("select * from retail where country = 'UK'");
ukTransactions.describe("unitPrice","quantity").show();
```

This will print the general stats such as count, mean, minimum, and maximum values on the columns unit price and quantity:

```
+-------+------------------+------------------+
|summary|         unitPrice|          quantity|
+-------+------------------+------------------+
|  count|            361878|            361878|
|   mean| 3.256006897353581|11.077028722387103|
| stddev|  70.6547311689391| 263.1292655989739|
|    min|               0.0|            -80995|
|    max|           38970.0|             80995|
+-------+------------------+------------------+
```

As you can see, the quantity values do not make much sense; this is probably due to bad data. The standard deviation in the quantity also shows that there are lots of bulk orders, which might lead us to the conclusion that most of the customers on these websites are corporate customers and not individuals.

Finally, there is one more important piece of information that we must check from data exploration for clustering, and that is checking if outliers are present in the data. Clustering, as such, is sensitive to the presence of outliers and in a big dataset like this we should discard any outliers or they might get clustered in a group of their own resulting in bad clustering. To figure out outliers, you can use different types of charts like box charts or histograms on the data. From the Spark side, you can use the `filter` function to filter out the outliers based on items like quantity, price, and so on. Clustering results will improve once the outliers are removed.

Let's now finally perform the clustering on this dataset and segment the customers into five different groups.

Clustering for customer segmentation

Here, we will now build a program that will use the k-means clustering algorithm and will make five clusters from our transactional dataset.

Before we crunch the data to figure out the clusters, we have made a few important assumptions and deductions regarding the data to preprocess it:

- We are only going to do clustering for the data belonging to the United Kingdom. The reason being, most of the data belongs to the United Kingdom in this dataset.
- For any missing or null values, we will simply discard that row of data. This is to keep things simple, and also because we have a good amount of data available for analysis. Leaving a few rows should not have much impact.

Let's now start our program. We will first build our boilerplate code to build the `SparkSession` and Spark configuration:

```
SparkConf conf = ...
SparkSession session = ...
```

Next, let's load the data from the file into a dataset:

```
Dataset<Row> rawData = session.read().csv("data/retail/Online_Retail.csv");
```

> We have used the `csv` method to read the data. This helps us read the `csv` data nicely, especially when the data has quotes in it. The `csv` reader package within Spark takes care of the double quotes, quotes, and so on, by default. We won't have to write any extract code for that.

As we mentioned in the data exploration phase, we are only going to run clustering on the data for the United Kingdom, so we will filter out the rows for United Kingdom as follows:

```
Dataset<Row> rawDataUK = rawData.filter("_c7 = 'United Kingdom'");
```

Next, we will convert this dataset into a Java RDD and fill it with a `RetailVO` POJO Java object. This POJO contains the following entries (to maintain brevity, we will show only the first few lines of code here):

```
public class RetailVO {
    private Integer quantity;
    private String invoiceDate;
    private Double unitPrice;
```

```
    ...
    private long amountSpent;
    private long recency;
```

As you can see, apart from the regular columns of the dataset that are mapped to this POJO (`RetailVO`), we have additional entries for the total amount spent by the customer, as well as the number of days that have passed since the customer made their last purchase.

We will invoke the mapper on the Java RDD and will tie a Java lambda function within it. We will read and extract the data from each row of the dataset and we will fill the data into an instance of a Java POJO called `RetailVO` (the code is the same as we showed previously):

```
JavaRDD<RetailVO> retailData = rawDataUK.javaRDD().map(row -> {
RetailVO retailVO = new RetailVO();
    String invoiceNo = row.getString(0);
    String stockCode = row.getString(1);
    String description = row.getString(2);
    ...
```

As you can see, we are pulling the value from the `row` object and storing it in the JavaBean. Next, we check if any value is `null` or missing; if it is, then we simply return `null` from this lambda function:

```
if( null == invoiceNo || "".equals(invoiceNo.trim()) ||
  null == stockCode || "".equals(stockCode.trim()) || ...) {
  return null;
}
```

We also convert the unit price to double and calculate the total amount spent (by multiplying the quantity purchased with the unit price of the item) and set it in the Java object:

```
Double unitPriceDbl = Double.parseDouble(unitPrice);
      retailVO.setUnitPrice( unitPriceDbl  );
   Integer quantityInt = Integer.parseInt(quantity);
      retailVO.setQuantity(quantityInt);
   long amountSpent = Math.round(unitPriceDbl * quantityInt );
      retailVO.setAmountSpent(amountSpent);
```

We also calculate the recency value here, that is, the days passed since the last purchase. Since this dataset is quite old, we are using 2012 as the year to keep the days count value reduced:

```
SimpleDateFormat myFormat = new SimpleDateFormat("MM/dd/yyyy");
Date date1 = myFormat.parse(invoiceDate);
Date date2 = myFormat.parse("12/31/2012");
long diff = date2.getTime() - date1.getTime();
long days = TimeUnit.DAYS.convert(diff, TimeUnit.MILLISECONDS);
        retailVO.setRecency(days);
    return retailVO;
}).filter( rowObj -> null != rowObj);
```

As you can see, we used the `SimpleDateFormat` class to figure out the recency value in days. In addition, from the RDD, we are filtering out all the null values, thus missing values in the dataset are simply discarded and not used in clustering.

Next, we will create a dataset out of this:

```
Dataset<Row> retailDS = session.createDataFrame(retailData.rdd(),
RetailVO.class);
```

We will also register this dataset as a temporary view to fire queries on this:

```
retailDS.createOrReplaceTempView("transactions");
```

We will fire the query to figure out the minimum value of the recency time per customer. We are only interested in the most recent purchase as per the RFM values we discussed earlier. We use the `min` function on the `transactions` view we created earlier and group the `transactions` by customer ID. We also register the resulting dataset as a temporary view called `recencyData`:

```
Dataset<Row> recencyDS = session.sql("select customerID,min(recency)
recency from
transactions group by customerID");
recencyDS.createOrReplaceTempView("recencyData");
```

Next, we find the frequency, which is the total count of transactions per customer. Again, we will run a `group` by query with a count on it. We will also register it as a temporary view called `freqData`:

```
Dataset<Row> freqDS = session.sql("select customerID,count(*)
frequency from transactions group by customerID");
freqDS.createOrReplaceTempView("freqData");
```

After finding the frequency, we find the monetary or the total amount spent by the customer on all their transactions. We will again run a `group by` query and sum up all the amounts spent by customers:

```
Dataset<Row> monetoryDS = session.sql("select
customerID,sum(amountSpent)
   spending from transactions group by customerID");
monetoryDS.createOrReplaceTempView("spendingData");
```

Now let's pull and combine all this data for recency, frequency, and monetary amount per customer into a single dataset. For this, we will fire a query on the views we created earlier and join them using the `customerID` field:

```
Dataset<Row> resultDS = session.sql("select r.customerID, r.recency,
f.frequency,
s.spending from recencyData r, freqData f, spendingData s where
r.customerID = f.customerID and f.customerID = s.customerID");
```

We can run a `show` method on this dataset, as follows:

```
resultDS.show();
```

This will print the following values:

customerID	recency	frequency	spending
12847	410	91	871
13192	483	63	922
13282	406	40	1055
13610	400	228	1129
13772	421	177	1145
13865	446	30	508
14157	407	49	401
14204	390	44	163
14887	467	6	1862
15269	411	2	409
15271	395	275	2507
15555	400	925	4784
15574	565	168	713
15634	405	15	246
16250	649	24	392
16504	413	86	486
17427	459	2	101
17506	458	16	297

As you can see the values differ a lot in terms of number. As by default clustering internally mostly uses the Euclidean distances the calculated similarity result might be impacted on as higher number values will get more weightage than others. So, to fix this issue we need to normalize these values.

Clustering and Customer Segmentation on Big Data

Before we normalize the values let's first extract these features and put them into a vector form and later we can normalize the values in this vector. To create a vector representation of these features we invoke the `VectorAssembler` class from the Spark `ml` package and provide the three columns from the previous dataset that have to be part of the vector. We also provide the resulting output column name and name it as `features`:

```
VectorAssembler assembler = new VectorAssembler().setInputCols(new String[]
{"recency","frequency","spending"}).setOutputCol("features");
```

We will run transform on the entire dataset and this will transform the three values (recency, frequency, and monetory) in vector format:

```
Dataset<Row> datasetWithFeatures = assembler.transform(resultDS);
```

Now, on these vectorized features we run our normalizer. There is a very nice normalizer bundled in Spark ML library. This normalizer would convert the output of the vector features into an output belonging to a similar scale. To this normalizer we supply the input data from the `features` column and it will print the normalized values in an output column:

```
Normalizer normalizer = new Normalizer().setInputCol("features").setOutputCol("normFeatures");
```

After forming the normalizer lets run it on the dataset that contains the features in vector form. The output of this transformation is also stored in a dataset. We will run a `show` method on this dataset to see the normalized values:

```
Dataset<Row> normDataset = normalizer.transform(datasetWithFeatures);
normDataset.show(10);
```

This will print the first ten normalized values as follows:

```
+----------+-------+---------+--------+-------------------+--------------------+
|customerID|recency|frequency|spending|           features|        normFeatures|
+----------+-------+---------+--------+-------------------+--------------------+
|     12847|    410|       91|     871| [410.0,91.0,871.0]|[0.42400691750858...|
|     13192|    483|       63|     922| [483.0,63.0,922.0]|[0.46319519866007...|
|     13282|    406|       40|    1055|[406.0,40.0,1055.0]|[0.35893227147767...|
|     13610|    400|      228|    1129|[400.0,228.0,1129.0]|[0.32806452843731...|
|     13772|    421|      177|    1145|[421.0,177.0,1145.0]|[0.34152159327630...|
|     13865|    446|       30|     508| [446.0,30.0,508.0]|[0.65911170105826...|
|     14157|    407|       49|     401| [407.0,49.0,401.0]|[0.70973271727211...|
```

As you can see the features are normalized and stored in a new vector under the column `normFeatures`.

We will also register this dataset as a temporary view and we will later use it in our analytics queries on the predicted clusters:

```
normDataset.createOrReplaceTempView("norm_data");
```

Now its time to build the actual clustering algorithm and we are using k-means algorithm. Create an instance of k-means which is available in the Spark ML package and supply the necessary parameters like the value of *k* (number of clusters) and the column where the features can be read, that is, `normFeatures` in our case:

```
KMeans km = new KMeans().setK(5).setSeed(1L).
setFeaturesCol("normFeatures");
```

Now fit the model on our dataset and this will generate a `KmeansModel` instance:

```
KMeansModel kmodel = km.fit(normDataset);
```

Finally run the transformation to collect the predicted clusters:

```
Dataset<Row> clusters = kmodel.transform(normDataset);
```

Let's see the first few rows of this predicted clusters dataset. We will invoke the `show` function on this dataset:

```
clusters.show()
```

This will print the first few rows as follows:

```
17/05/17 17:02:24 INFO CodeGenerator: Code generated in 11.46881 ms
+----------+-------+---------+--------+--------------------+--------------------+----------+
|customerID|recency|frequency|spending|            features|        normFeatures|prediction|
+----------+-------+---------+--------+--------------------+--------------------+----------+
|     12847|    410|       91|     871|  [410.0,91.0,871.0]|[0.42400691750858...|         3|
|     13192|    483|       63|     922|  [483.0,63.0,922.0]|[0.46319519866007...|         3|
|     13282|    406|       40|    1055| [406.0,40.0,1055.0]|[0.35893227147767...|         3|
|     13610|    400|      228|    1129|[400.0,228.0,1129.0]|[0.32806452843731...|         3|
|     13772|    421|      177|    1145|[421.0,177.0,1145.0]|[0.34152159327630...|         3|
|     13865|    446|       30|     508|  [446.0,30.0,508.0]|[0.65911170105826...|         1|
|     14157|    407|       49|     401|  [407.0,49.0,401.0]|[0.70973271727211...|         1|
|     14204|    390|       44|     163|  [390.0,44.0,163.0]|[0.91769786703235...|         4|
|     14887|    467|        6|    1862|  [467.0,6.0,1862.0]|[0.24326978761296...|         2|
```

As you can see the predicted clusters are named from zero to four (hence five clusters). The predicted cluster number is stored in the `prediction` column.

We will register the predicted dataset as a temporary view called `cluster` to further analyze the clustered results:

```
clusters.createOrReplaceTempView("clusters");
```

Let's now find how many items are assigned per cluster in our dataset. To do this, we will run a group by query on our view for `clusters` and do a count of items per cluster:

```
session.sql("select prediction,count(*) from clusters group by
prediction").show();
```

This will print the result as follows:

```
+----------+--------+
|prediction|count(1)|
+----------+--------+
|         1|     668|
|         3|     822|
|         4|     716|
|         2|    1133|
|         0|     611|
+----------+--------+
```

As you can see our groups are nicely distributed.

> As you can see the number of total customers in the groups is very low. This again gives the conclusion that most of these are corporate customers who are making large transactions on the website.

Now let's try to find who the valuable customers, are for the online store and let's check the group they fall into. To do this we will fire a query and search on group 2. We would suggest that the reader does simple analytics on other groups as well:

```
session.sql("select * from norm_data where customerID in (select
customerID from
clusters where prediction = 2)").show();
```

```
17/05/17 22:45:18 INFO DAGScheduler: Job 43 finished: show at KMeansClustering.java:139, took 0.127328 s
+----------+-------+---------+--------+--------------------+--------------------+
|customerID|recency|frequency|spending|            features|        normFeatures|
+----------+-------+---------+--------+--------------------+--------------------+
|     14887|    467|        6|    1862| [467.0,6.0,1862.0]|[0.24326978761296...|
|     15271|    395|      275|    2507|[395.0,275.0,2507.0]|[0.15473311361757...|
|     15555|    400|      925|    4784|[400.0,925.0,4784.0]|[0.08181639081528...|
|     17686|    395|      286|    5786|[395.0,286.0,5786.0]|[0.06802703458963...|
|     13985|    392|      353|    7072|[392.0,353.0,7072.0]|[0.05527629909846...|
|     15947|    470|       29|    1709| [470.0,29.0,1709.0]|[0.26513411031107...|
|     16549|    398|      981|    4200|[398.0,981.0,4200.0]|[0.09188778871879...|
|     17757|    389|      742|    5645|[389.0,742.0,5645.0]|[0.06816393548275...|
|     13107|    432|       60|    1526| [432.0,60.0,1526.0]|[0.27219383443550...|
|     14525|    396|      298|    4242|[396.0,298.0,4242.0]|[0.09272152767339...|
|     15478|    428|       46|    1449| [428.0,46.0,1449.0]|[0.28314576816739...|
|     16303|    413|      167|    5313|[413.0,167.0,5313.0]|[0.07746203711191...|
|     18283|    391|      756|    2140|[391.0,756.0,2140.0]|[0.16977522070462...|
```

As you can see this group contains the customers who have spent the highest amounts. These customers are valuable as they spend a lot as compared to other customers (you can check this by analyzing the other groups similarly).

With this we have built a customer segmentation solution for our retail dataset using k-means clustering. But what if we want to change the algorithm and use bisecting k-means clustering.

Changing the clustering algorithm

We mentioned the bisecting k-means algorithm. This algorithm is bundled in the Spark ML library. To change our regular k-means algorithm to this algorithm and run the customer segmentation code with it, we just need to change the portion of the code where we create the instance of the k-means algorithm and apply it on the dataset to make the cluster predictions. All remaining code remains the same. So the changed code for the algorithm will have the usual instance creation of the algorithm followed by setting the parameters like the number of clusters that we are looking for and the column from where the algorithm can read the input features. Later we will again fit and transform the model on the dataset:

```
BisectingKMeans bkm = new
BisectingKMeans().setK(5).setSeed(1L).setFeaturesCol("normFeatures");
BisectingKMeansModel bkmodel = bkm.fit(normDataset);
Dataset<Row> clusters = bkmodel.transform(normDataset);
```

Now if we run our query for counting the items in each cluster we will get the results as follows:

```
+----------+--------+
|prediction|count(1)|
+----------+--------+
|         1|     601|
|         3|     833|
|         4|     926|
|         2|     917|
|         0|     673|
+----------+--------+
```

You will notice that these results are much better than k-means simply because bisecting k-means is a better performer than plain k-means algorithm.

We can also check some of the items present in the cluster 2 and this will print some items from it as follows:

```
17/05/18 00:05:12 INFO DAGScheduler: Job 79 finished: show at BisectingKmeansClustering.java::
+----------+-------+---------+--------+------------------+--------------------+
|customerID|recency|frequency|spending|          features|        normFeatures|
+----------+-------+---------+--------+------------------+--------------------+
|     12847|    410|       91|     871| [410.0,91.0,871.0]|[0.42400691750858...|
|     13192|    483|       63|     922| [483.0,63.0,922.0]|[0.46319519866007...|
|     13282|    406|       40|    1055|[406.0,40.0,1055.0]|[0.35893227147767...|
|     13865|    446|       30|     508| [446.0,30.0,508.0]|[0.65911170105826...|
|     15574|    565|      168|     713|[565.0,168.0,713.0]|[0.61074193335466...|
|     16504|    413|       86|     486| [413.0,86.0,486.0]|[0.64174937053506...|
|     15539|    395|       41|     539| [395.0,41.0,539.0]|[0.58999467069230...|
|     16027|    479|       17|     851| [479.0,17.0,851.0]|[0.49043000426939...|
|     16340|    495|      153|     563|[495.0,153.0,563.0]|[0.64696044239096...|
|     13122|    482|       55|     929| [482.0,55.0,929.0]|[0.45990565858249...|
```

If you look closely at these results and cross compare them with the results we got for our results with plain k-means you would find that these results are different than normal k-means.

With this we have covered our case studies in this chapter. We would mention to the users that we have just scratched the surface of clustering in this chapter and it is a very wide topic and entire books have been written on it. We would urge the users to research more on this topic and practice on different datasets.

Summary

In this chapter, we learnt about clustering and we saw how this approach helps to group different items into groups with each group having items which are similar to them in some form. Clustering is an example of unsupervised learning and there are lots of popular clustering algorithms that are shipped by default in the Apache Spark package. We learnt about two clustering approaches, the first being k-means approach where items that are closer to each other based on some mathematical formula like Euclidean distance and so on were grouped together. We also learnt about bisecting k-means approach which is essentially and improvement on the regular k-means clustering and is creating by being a combination of hierarchical and k-means clustering. We also applied clustering on a sample dataset of retail from UCI. On this sample case study we segmented the customers of the website using clustering and tried to figure out the important customers for an online e-commerce store.

In the next chapter, we will learn about an important and powerful concept from our computer science and that is graph and their application in analytics using big data.

11
Massive Graphs on Big Data

Graph theory is one of the most important and interesting concepts in computer science. Graphs have been implemented in real life in a lot of use cases. If you use a GPS on your phone or a GPS device and it shows you driving directions to a place, behind the scenes, there is an efficient graph that is working for you to give you the best possible directions. In a social network, you are connected to your friends and your friends are connected to other friends, and so on. This is a massive graph running in production in all the social networks that you use. You can send messages to your friends, follow them, or get followed, all in this graph. Social networks or a database storing driving directions all involve massive amounts of data, and this is not data that can be stored on a single machine; instead, this is distributed across a cluster of thousands of nodes or machines. This massive data is nothing but big data and, in this chapter, we will learn how data can be represented in the form of a graph so that we can perform analysis or deductions on top of these massive graphs.

In this chapter, we will cover:

- A short refresher on graphs and their basic concepts
- A short introduction to the GraphStream library for representing graph charts.
- A short introduction to graph analytics, its advantages, and how Apache Spark fits in
- An introduction to GraphFrames library that is used on top of Apache Spark
- Building graphs using GraphFrames library
- Case study on graph analytics on top of airports and their flights data.
- Finding the top airports using the PageRank algorithm

Before we dive deeply into each individual section, let's look at the basic graph concepts in brief.

Refresher on graphs

In this section, we will cover some of the basic concepts of graphs; this is supposed to be a refresher section on graphs. This is a basic section; hence, if you already know this information, you can skip this section. Graphs are used in many important concepts in our day-to-day lives. Before we dive into the ways of representing a graph, let's look at some of the popular use cases of graphs (though this is not a complete list):

- Graphs are used heavily in social networks
- In finding driving directions via GPS
- In many recommendation engines
- In fraud detection in many financial companies
- In search engines and in network traffic flows
- In biological analysis

As you must have noted earlier, graphs are used in many applications that we might be using on a daily basis.

Graphs are a form of a data structure in computer science that help in depicting entities and the connection between them. So, if there are two entities, such as Airport A and Airport B and they are connected by a flight that takes, for example, a few hours, then Airport A and Airport B are the two entities and the flight connecting them that takes those specific hours depicts the weightage between them or their connection. In formal terms, these entities are called **vertexes** and the relationships between them are called **edges**. So, in mathematical terms, a graph $G = \{V, E\}$; that is, a graph is a function of vertexes and edges. Let's look at the following diagram for a simple example of a graph:

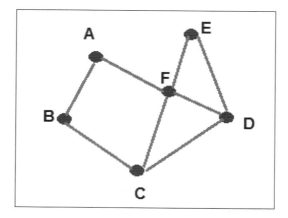

As you can see, the preceding graph is a set of six vertexes and eight edges, as shown next:

Vertexes = {A, B, C, D, E, F}

Edges = {AB, AF, BC, CD, CF, DF, DE, EF}

These vertexes can represent any entities, for example, they can be places, with the edges being distances between the places; or they could be people in a social network, with the edges being the type of relationship, for example, friends or followers. Thus, graphs can represent real-world entities like this.

The preceding graph is also called a **bidirected** graph because, in this graph, the edges go in either direction, that is, the edge from **A** to **B**, can be traversed both ways from **A** to **B** as well as from **B** to **A**. Thus, the edge in the preceding diagram that is **AB** can be **BA**, and **AF** can be **FA** too. There are other types of graphs, called directed graphs, and in these graphs, the direction of the edges goes in one way only and does not retrace back. A simple example of a directed graph is shown as follows:

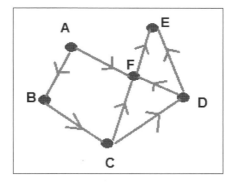

As seen in the preceding graph, the edge **A** to **B** goes only in one direction, as does the edge **B** to **C**. Hence, this is a directed graph.

 A simple linked list data structure, or a tree data structure, are also forms of graph. In a tree, nodes can only have children and there are no loops; while there is no such rule in a general graph.

Representing graphs

Visualizing a graph makes it easily comprehensible, but depicting it using a program requires two different approaches:

- **Adjacency matrix**: Representing a graph as a matrix is easy and has its own advantages and disadvantages. Let's look at the bidirected graph that we showed in the preceding diagram. If you represented this graph as a matrix, it would look like this:

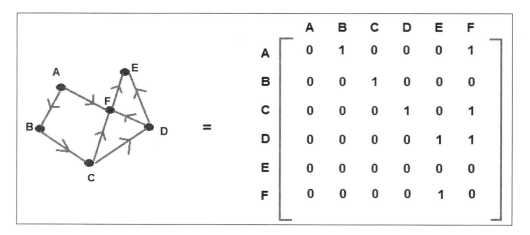

The preceding diagram is a simple representation of our graph in matrix form. The concept of matrix representation of a graph is simple—if there is an edge to a node, we mark the value as 1, else, if the edge is not present, we mark it as 0. As this is a bi-directed graph, it has edges flowing in one direction only. Thus, from the matrix, the rows and columns depict the vertices. There, if you look at the vertex **A**, it has an edge to vertex **B**, and the corresponding matrix value is 1.

As you can see, it takes just one step or O[1] to figure out an edge between two nodes. We just need the index (rows and columns) in the matrix and we can extract that value from it. Also, if you looked at the matrix closely, you would have seen that most of the entries are zero, hence this is a sparse matrix. Thus, this approach eats a lot of space in computer memory in marking even those elements that do not have an edge to each other, and this is its main disadvantage:

- **Adjacency list**: An adjacency list solves the problem of space wastage of the adjacency matrix. To solve this problem, it stores the node and its neighbors in a list (linked list), as shown in the following diagram:

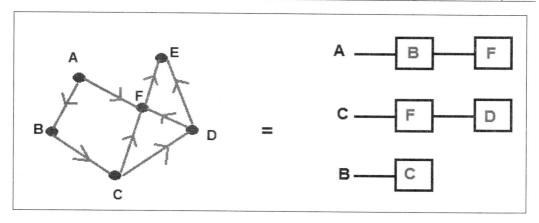

To maintain brevity, we have not shown all the vertices, but you can make out from the diagram that each vertex stores its neighbors in a linked list. So, when you want to figure out the neighbors of a particular vertex, you can directly iterate over the list. Of course, this has the disadvantage of iterating when you have to figure out whether an edge exists between two nodes or not. This approach is also widely used in many algorithms in computer science.

We have briefly looked at how graphs can be represented, let's now look at some important terms that are used heavily on graphs.

Common terminology on graphs

We will now introduce you to some common terms and concepts in graphs that you can use in your analytics on top of graphs:

- **Vertices**: As we mentioned earlier, vertices are the mathematical terms for the nodes in a graph. For analytic purposes, the vertices count shows the number of nodes in the system, for example, the number of people involved in a social graph.
- **Edges**: As we mentioned earlier, edges are the connection between vertices and edges can carry weights. The number of edges represents the number of relations in a graph system. The weight on a graph represents the intensity of the relationship between the nodes involved; for example, in a social network, the relationship of friends is a stronger relationship than followers between nodes.

- **Degrees**: Represent the total number of connections flowing into, as well as out of a node. For example, in the previous diagram, the degree of node **F** is **four**. The degree count is useful, for example, in a social network graph it can represent how well a person is connected if his degree count is very high.
- **Indegrees**: This represents the number of connections flowing into a node. For example, in the previous diagram, for node **F**, the indegree value is three. In a social network graph, this might represent how many people can send messages to this person or node.
- **Outdegrees**: This represents the number of connections flowing out of a node. For example, in the previous diagram, for node **F**, the outdegree value is one. In a social network graph, this might represent how many people can send messages to this person or node.

Common algorithms on graphs

Let's look at the three common algorithms that are run on graphs frequently and some of their uses:

- **Breadth first search**: Breadth first search is an algorithm for graph traversal or searching. As the name suggests, the traversal occurs across the breadth of the graphs, that is to say, the neighbors of the node from where traversal starts are searched first, before exploring further in the same manner. We will refer to the same graph we used earlier:

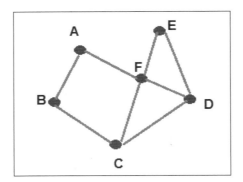

If we start at vertex **A**, then, according to a breadth first search, next, we search or go to the neighbors of **A**, that is, **B** and **F**. After that, we will go to the neighbor of **B**, and that will be **C**. Next, we will go to the neighbors of **F**, and those will be **E** and **D**. We only go through each node once, and this mimics real-life travel as well, as to reach a point from another point, we seldom cover the same road or path again.

Thus, our breadth first traversal starting from **A** will be {A , B , F , C , D , E }.

Breadth first search is very useful in graph analytics and can tell us things such as the friends that are not your immediate friends but just at the next level after your immediate friends, in a social network; or, in the case of a graph of a flights network, it can show flights with just a single stop or two stops to the destination.

- **Depth-first search**: This is another way of searching, where we start from the source vertices and keep on searching until we reach the end node or the leaf node, and then we backtrack. This algorithm is not as performant as the breadth first search, as it requires lots of traversals. So, if you want to know whether node **A** is connected to node **B**, you might end up searching along a lot of wasteful nodes that do not have anything to do with the original nodes **A** and **B** before coming to the appropriate solution.

- **Dijkstra's shortest path**: This is a greedy algorithm to find the shortest path in a graph network. So, in a weighted graph, if you need to find the shortest path between two nodes, you can start from the starting node and keep on picking the next node in the path greedily to be the one with the least weight (in the case of weights being distances between nodes, as in city graphs depicting interconnecting cities and roads). So, in a road network, you can find the shortest path between two cities using this algorithm.

- **PageRank algorithm**: This is a very popular algorithm that came out from Google and is essentially used to find the importance of a web page by figuring out how connected it is to other important websites. It gives a page rank score to each website based on this approach and, finally, the search results are built based on this score. The best part about this algorithm is it can be applied to other areas of life too, for example, in figuring out the important airports on a flight graph, or figuring out the most important people in a social network group.

Enough the basics and the refresher on graphs. In the next section, we will see how graphs can be used in the real world in massive datasets, such as social network data, or in data used in the field of biology. We will also study how graph analytics can be used on top of these graphs to derive exclusive deductions.

Plotting graphs

There is a handy open source Java library called GraphStream, which can be used to plot graphs, and this is very useful, especially if you want to view the structure of your graphs. While viewing, you can also figure out whether some of the vertices are very close to each other (clustered), or in general, how they are placed.

Using the GraphStream library is easy. Just download the jar from http://graphstream-project.org and put it in the classpath of your project. Next, we will show a simple example demonstrating how easy it is to plot a graph using this library.

> This library is extensive and it will be a good learning experience to explore this library further. We would urge readers to further explore this library on their own.

Just create an instance of a graph. For our example, we will create a simple `DefaultGraph` and name it `SimpleGraph`. Next, we will add the nodes or vertices of the graph. We will also add the attribute of the label that is displayed on the vertice:

```
Graph graph = newDefaultGraph("SimpleGraph");
  graph.addNode("A" ).setAttribute("ui.label", "A");
  graph.addNode("B" ).setAttribute("ui.label", "B");
  graph.addNode("C" ).setAttribute("ui.label", "C");
```

After building the nodes, it's now time to connect these nodes using the edges. The API is simple to use and on the graph instance we can define the edges, provided an ID is given to them and the starting and ending nodes are also given:

```
graph.addEdge("AB", "A", "B");
graph.addEdge("BC", "B", "C");
graph.addEdge("CA", "C", "A");
```

All the information for nodes and edges is present on the graph instance. It's now time to plot this graph on the UI, and we can just invoke the display method on the graph instance, as shown next, and display it on the UI:

```
graph.display();
```

This will plot the graph on the UI as follows:

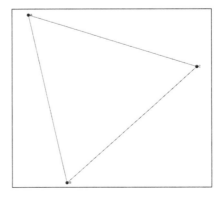

Chapter 11

Massive graphs on big data

Big data comprises a huge amount of data distributed across a cluster of thousands (if not more) of machines. Building graphs based on this massive data has different challenges. Due to the vast amount of data involved, the data for the graph is distributed across a cluster of machines. Hence, in actuality, it's not a single node graph, and we have to build a graph that spans across a cluster of machines. A graph that spans across a cluster of machines would have vertices and edges spread across different machines, and this data in a graph won't fit into the memory of one single machine. Consider your friend's list on Facebook; some of your friend's data in your Facebook friend list graph might lie on different machines, and this data may just be tremendous in size. Look at an example diagram of a graph of 10 Facebook friends and their network, shown as follows:

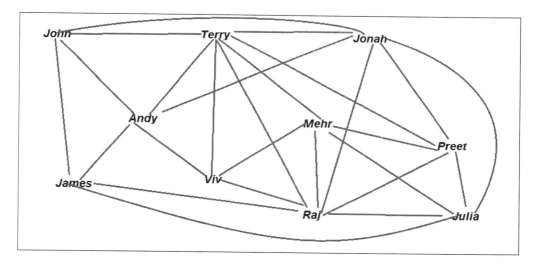

As you can see in the preceding diagram, for just 10 friends the data can be huge, and here, since the graph was drawn by hand, we have not even shown a lot of connections to make the image comprehensible. In real life, each person can have, say, thousands of connections. So, imagine what would happen to a graph with thousands, if not more, people on the list.

For the reasons we have just seen, building massive graphs on big data is a different ball game altogether, and there are a few main approaches to building these massive graphs. From the perspective of big data, building massive graphs involves running and storing data parallely on many nodes. The two main approaches are bulk synchronous parallely and the Pregel approach. Apache Spark follows the Pregel approach. Covering these approaches in detail is out of the scope of this book. If you are interested in these topics, you should refer to other books and the Wikipedia page for the same.

Graph analytics

The biggest advantage to using graphs is you can analyze those graphs and use them to analyze complex datasets. You might ask, what is so special about graph analytics that we can't do using relational databases. Let's try to understand this using an example. Suppose we want to analyze your friends network on Facebook, and pull information about your friends such as their name, their birth date, their recent likes, and so on. If Facebook had a relational database, then this would mean firing a query on a table using the foreign key of the user requesting this info. From the perspective of a relational database, this first level query is easy. But what if we now ask you to go to the friends at level four in your network and fetch their data (as shown in the following diagram). The query to get this becomes more and more complicated from a relational database perspective, but this is a trivial task on a graph or graphical database (such as Neo4j). Graphs are extremely good in operations where you want to pull information from one end of the node to another, where the other node lies after a lot of joins and hops. As such, graph analytics is good for certain use cases (but not for all use cases; relational databases are still good in many other use cases):

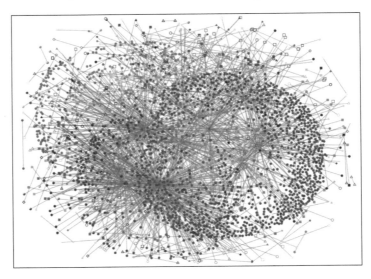

As you can see, the preceding diagram depicts a huge social network (though it might just be depicting a network of a few friends). The dots represent actual people in a social network. So, if somebody asks to pick one user on the left most side of the diagram, see and follow host connections to the right most side and pull the friends at the, say, 10th level or more, this would be something very difficult to do in a normal relational database and doing it and maintaining it could easily get out of hand.

There are four particular use cases where graph analytics is extremely useful and used frequently (though there are plenty more use cases too):

- **Path analytics**: As the name suggests, this analytics approach is used to figure out the paths as you traverse along the nodes of a graph. There are many fields where this can be used—the simplest being road networks and figuring out details such as the shortest path between cities, or in flight analytics to figure out the flight taking the shortest time or direct flights.

- **Connectivity analytics**: As the name suggests, this approach outlines how the nodes within a graph are connected to each other. So, using this, you can figure out how many edges are flowing into a node and how many are flowing out of the node. This kind of information is very useful in analysis. For example, in a social network, if there is a person who receives just one message, but gives out, say, ten messages within his network then this person can be used to market his favorite products, as he is very good at responding to messages.

- **Community Analytics**: Some graphs on big data are huge. But, within these huge graphs, there might be nodes that are very close to each other and are almost stacked in a cluster of their own. This is useful information as based on this you can extract communities from your data. For example, in a social network if there are people who are part of a community, say marathon runners, then they can be clubbed into a single community and further tracked.

- **Centrality Analytics**: This kind of analytical approach is useful in finding central nodes in a network or graph. This is useful in figuring out sources that are single-handedly connected to many other sources. It is helpful in figuring out influential people in a social network, or a central computer in a computer network.

From the perspective of this chapter, we will be covering some of these use cases in our sample case studies, and for this, we will be using a library on Apache Spark called GraphFrames.

GraphFrames

The GraphX library is advanced and performs well on massive graphs, but, unfortunately, it's currently only implemented in Scala and does not have any direct Java API. GraphFrames is a relatively new library that is built on top of Apache Spark and provides support for dataframe (now dataset) based graphs. It contains a lot of methods that are direct wrappers over the underlying `sparkx` methods. As such, it provides similar functionality to GraphX, except that GraphX acts on the Spark RDD and GraphFrame works on the dataframe, so GraphFrame is more user-friendly (as dataframes are simpler to use). All the advantages of firing Spark SQL queries, joining datasets, and filtering queries are supported on this.

To understand GraphFrames and represent massive big data graphs, we will take baby steps first by building some simple programs using GraphFrames, before building full-fledged case studies. First, let's see how to build a graph using Spark and GraphFrames on a sample dataset.

Building a graph using GraphFrames

Consider that you have a simple graph as shown next. This graph depicts four people **Kai**, **John**, **Tina**, and **Alex** and the relationships they share, whether they follow each other or are friends.

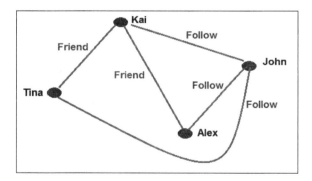

We will now try to represent this basic graph using the GraphFrame library on top of Apache Spark and, in the meantime, we will also start learning the GraphFrame API.

Since GraphFrame is a module on top of Spark, let's first build the Spark configuration and `spark sql` context for brevity:

```
SparkConfconf= ...
JavaSparkContextsc= ...
SQLContextsqlContext= ...
```

We will now build the `JavaRDD` object that will contain the data for our vertices, or the people `Kai`, `John`, `Alex`, and `Tina`, in this small network. We will create some sample data using the `RowFactory` class of the Spark API and provide the attributes (ID of the person, and their name and age) that we need per row of the data:

```
JavaRDD<Row>verRow =
sc.parallelize(Arrays.asList(RowFactory.create(101L,"Kai",27),
     RowFactory.create(201L,"John",45),
     RowFactory.create(301L,"Alex",32),
     RowFactory.create(401L,"Tina",23)));
```

Next, we will define the structure or schema of the attributes used to build the data. The ID of the person is of type long, the name of the person is a string, and the age of the person is an integer, as shown next in the code:

```
List<StructField>verFields = newArrayList<StructField>();
   verFields.add(DataTypes.createStructField("id",DataTypes.LongType,
true));
   verFields.add(DataTypes.createStructField("name",DataTypes.
StringType,
true));
       verFields.add(DataTypes.createStructField("age",DataTypes.
IntegerType, true));
```

Now, let's build the sample data for the relationships between these people, and this will, basically, be represented as the edges of the graph later. The data item of relationship will have the IDs of the persons that are connected together and the type of relationship they share (that is, friends or followers). Again, we will use the Spark-provided `RowFactory`, build some sample data per row, and create the `JavaRDD` with this data:

```
JavaRDD<Row>edgRow = sc.parallelize(Arrays.asList(
     RowFactory.create(101L,301L,"Friends"),
     RowFactory.create(101L,401L,"Friends"),
     RowFactory.create(401L,201L,"Follow"),
     RowFactory.create(301L,201L,"Follow"),
     RowFactory.create(201L,101L,"Follow")));
```

Again, define the schema of the attributes added as part of the edges earlier. This schema is later used in building the dataset for the edges. The attributes passed are the source ID of the node, destination ID of the other node, as well as the `relationType`, which is a string:

```
List<StructField>EdgFields = newArrayList<StructField>();
  EdgFields.add(DataTypes.createStructField("src",DataTypes.
LongType,true));
  EdgFields.add(DataTypes.createStructField("dst",DataTypes.
LongType,true));
  EdgFields.add(DataTypes.createStructField("relationType",DataTypes.
StringType,true));
```

Using the schemas that we have defined for the vertices and edges, let's now build the actual dataset for the vertices and the edges. To do this, first create the `StructType` object that holds the schema details for the vertices and the edges data and, using this structure and the actual data, we will next build the dataset of the vertices (`verDF`) and the dataset for the edges (`edgDF`):

```
StructTypeverSchema = DataTypes.createStructType(verFields);
StructTypeedgSchema = DataTypes.createStructType(EdgFields);

Dataset<Row>verDF = sqlContext.createDataFrame(verRow, verSchema);
Dataset<Row>edgDF = sqlContext.createDataFrame(edgRow, edgSchema);
```

Finally, we will now use the vertices and the edges dataset, pass it as a parameter to the `GraphFrame` constructor, and build the `GraphFrame` instance:

```
GraphFrameg = newGraphFrame(verDF,edgDF);
```

The time has now come to look at some mild analytics on the graph we have just created.

Let's first visualize our data for the graphs; let's look at the data on the vertices. For this, we will invoke the `vertices` method on the GraphFrame instance and invoke the standard `show` method on the generated vertices dataset (GraphFrame will generate a new dataset when the `vertices` method is invoked).

```
g.vertices().show();
```

Chapter 11

This will print the output as follows:

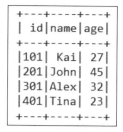

Let's also look at the data on the edges:

```
g.edges().show();
```

This will print the output as follows:

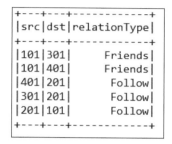

Let's also look at the number of edges and the number of vertices:

```
System.out.println("Number of Vertices : " + g.vertices().count());
System.out.println("Number of Edges : " + g.edges().count());
```

This will print the result as follows:

```
Number of Vertices : 4
Number of Edges : 5
```

GraphFrame has a handy method to find all the indegrees (outdegrees or degrees):

```
g.inDegrees().show();
```

This will print the indegrees of all the vertices, as shown next:

```
+---+--------+
| id|inDegree|
+---+--------+
|201|       2|
|301|       1|
|101|       1|
|401|       1|
+---+--------+
```

Finally, let's look at one more small thing on this simple graph. As GraphFrames works on the datasets, all the dataset handy methods, such as filtering, map, and so on can be applied to them. We will use the filter method and run it on the vertices dataset to figure out the people in the graph with an age greater than thirty:

```
g.vertices().filter("age > 30").show();
```

This will print the result as follows:

```
+---+----+---+
| id|name|age|
+---+----+---+
|201|John| 45|
|301|Alex| 32|
+---+----+---+
```

Well, enough of basic graphs using GraphFrames. Now let's get into some realistic use cases for graph analytics.

Graph analytics on airports and their flights

As is evident from the name of this section, the airports are vertices and the flights connecting them are their edges, and these entities can easily be put on a graph. Once we have built this graph, we can start analyzing it.

For our analysis in this section, we will be using many datasets, and they are explained in the next section.

Datasets

As part of the analysis done for this case study, we will be using the following datasets, as shown next. All these datasets have been obtained from the OpenFlights airports database (`https://openflights.org/data.html`).

- **Airports dataset**: This dataset contains information about the various airports used in this case study. There are many attributes but we only use a subset of them, hence, we will explain only the ones that we use next. We urge you to check out the other attributes, such as, latitude and longitude, further in your analysis:

Attribute	Description
`Airport ID`	This is the ID given to the airports per row in this dataset
`Airport IATA code`	This is the three-letter IATA (International Air Transport Association) airport code, which is null if not assigned/unknown
`Airport ICAO code`	This is the four-letter ICAO (International Civil Aviation Organization) airport code, which is null if not assigned
`Airport name`	This is the name of the airport
`Country`	This is the country in which the airport is located
`State`	This is the state in which the airport is located

- **Routes dataset**: This dataset contains information about the routes between the airports mentioned earlier. The attributes that we use from this dataset are:

Attribute	Description
`Airline`	This is the two-letter (IATA) or three-letter (ICAO) code of the airline
`Airline ID`	This is the unique OpenFlights identifier for the airline
`Source airport`	This is the three-letter (IATA) or four-letter (ICAO) code of the source airport
`Source airport ID`	This is the unique OpenFlights identifier for the source airport
`Destination airport`	This is the three-letter (IATA) or four-letter (ICAO) code of the destination airport
`Destination airport ID`	This is the unique OpenFlights identifier for the destination airport

- **Airlines dataset**: This dataset contains information about the airlines that are represented in this dataset. The attributes present in this dataset are:

Attribute	Description
`Airline ID`	This is the unique OpenFlights identifier for this airline
`Name`	This is the name of the airline
`IATA`	This is the two-letter IATA code, if available
`ICAO`	This is the three-letter ICAO code, if available
`Country`	This is the country or territory where the airline is incorporated.

We will now start analyzing these datasets in the next section.

Graph analytics on flights data

Before we run any analyses, we will build our regular Spark boilerplate code to get started. We will create the `SparkSession` to start loading our datasets:

```
SparkConfconf = ...
SparkSession session = ...
```

Once our `SparkSession` object is built, next, we will load our `airports` dataset. Since this dataset is in CSV format, we will load it with the CSV package present inside the Spark API. This will create a dataset of row objects. We will also print the first few values of this dataset to see how the data looks:

```
Dataset<Row>rawDataAirport = session.read().csv("data/flight/airports.dat");
    rawData.show();
```

This will print the data shown as shown in the succeeding image. The data shows the name of the airports, as well as in which country and state they are located:

```
+---+--------------------+------------------+----------------+---+----+-------------------+-------------------+----+
|_c0|                 _c1|               _c2|             _c3|_c4| _c5|                _c6|                _c7| _c8|
+---+--------------------+------------------+----------------+---+----+-------------------+-------------------+----+
|  1|      Goroka Airport|            Goroka|Papua New Guinea|GKA|AYGA| -6.081689834590001|    145.391998291|5282|
|  2|      Madang Airport|            Madang|Papua New Guinea|MAG|AYMD|      -5.20707988739|    145.789001465|  20|
|  3|Mount Hagen Kagam...|       Mount Hagen|Papua New Guinea|HGU|AYMH| -5.826789855957031|  144.29600524902344|5388|
|  4|      Nadzab Airport|            Nadzab|Papua New Guinea|LAE|AYNZ|          -6.569803|        146.725977| 239|
|  5|Port Moresby Jack...|      Port Moresby|Papua New Guinea|POM|AYPY| -9.443380355834961|  147.22000122070312| 146|
|  6|Wewak Internation...|             Wewak|Papua New Guinea|WWK|AYWK|       -3.58383011818|      143.669006348|  19|
|  7|   Narsarsuaq Airport|         Narssarssuaq|       Greenland|UAK|BGBW|       61.1604995728|      -45.4259986877| 112|
|  8| Godthaab / Nuuk A...|          Godthaab|       Greenland|GOH|BGGH|         64.19090271|      -51.6781005859| 283|
+---+--------------------+------------------+----------------+---+----+-------------------+-------------------+----+
```

Chapter 11

As you can see in the preceding screenshot, the data has default columns. Next, we will load each row of data into a JavaBean object and create an RDD of these Java objects. For brevity, we will not show the entire code here, but you can see that we are extracting the data from each row of the dataset and putting it into an `Airport` Java object. The main fields in this Java object (that you can see in our GitHub repository) are the `AirportID`, country, and the airport name:

```
JavaRDD<Airport>airportsRdd = 
        rawDataAirport.javaRDD().map(row -> {
          Airport ap = newAirport();
            ap.setAirportId(row.getString(0));
            ap.setState(row.getString(2));
            ...
          returnap;
        });
```

Since GraphFrame uses dataframes (or datasets), we convert this Java RDD into a dataset object. As this dataset has airports data that we will be later using in our analytical queries, we will also register this dataset as a temporary view:

```
Dataset<Row>airports = session.createDataFrame(airportsRdd.
rdd(),Airport.class);
airports.createOrReplaceTempView("airports");
```

> The `airports` dataset object we just saw is nothing but our dataset of vertices that we will use in our graph.

After loading the data for the airports, it's now time to load the data for the routes, and these routes will be the edges of our graph. The data is loaded from the `routes.dat` file (which is in CSV format) and loaded into a `Dataset` object:

```
Dataset<Row>rawDataRoute = session.read().csv("data/flight/routes.
dat");
```

Again, we load this routes data into a `Route` JavaBean. This JavaBean has attributes such as the airport ID from where the flight originates (that is, the `Src` attribute) on the airport ID of the flight's destination (that is, the 'dest' attribute).

```
JavaRDD<Route>routesRdd = 
      rawDataRoute.javaRDD().map(row -> {
        Route r = newRoute();
          r.setSrc(String)
          r.setDst(String)
  ...
      returnr;
    });
```

We will convert this `routesRdd` object to a dataset, which will serve as the dataset of our edges for the graph:

```
Dataset<Row>routes = session.createDataFrame(routesRdd.rdd(), Route.class);
```

Now, let's see the first few attributes of our route's dataset by invoking the `show` method on the routes dataset object.

```
routes.show();
```

This will print the output as shown here (it will print the source node and the destination node codes):

```
+-----------+---------+---+-------+-----+------------+---+-------+-----+
|airLineCode|airlineId|dst|dstCode|dstId|relationType|src|srcCode|srcId|
+-----------+---------+---+-------+-----+------------+---+-------+-----+
|        410|     null|KZN|   null| null|        null|AER|   null| null|
|        410|     null|KZN|   null| null|        null|ASF|   null| null|
|        410|     null|MRV|   null| null|        null|ASF|   null| null|
|        410|     null|KZN|   null| null|        null|CEK|   null| null|
+-----------+---------+---+-------+-----+------------+---+-------+-----+
```

It's now time to build our `GraphFramegf` object by providing the vertices dataset and the edges dataset (that is, `airports` and the `routes`):

```
GraphFramegf = newGraphFrame(airports, routes);
```

> It is important that you have an ID attribute in your vertice and a corresponding `Src` and `dest` attribute in your edge depicting object. These attributes are used internally by GraphFrames to build the graph object.

Now you have your `GraphFramegf` object ready. Let's now quickly see whether the object is good or not. To do this, we will just pull the vertices from this object and print the first few rows of the vertices in our graph:

```
gf.vertices().show();
```

This will print the vertices (and their attributes), as shown in the following output:

```
+-------------+--------------+---------+--------------------+----------------+---+
|airportIataCode|airportIcaoCode|airportId|         airportName|         country| id|
+-------------+--------------+---------+--------------------+----------------+---+
|          GKA|          AYGA|        1|     Goroka Airport|Papua New Guinea|GKA|
|          MAG|          AYMD|        2|     Madang Airport|Papua New Guinea|MAG|
|          HGU|          AYMH|        3|Mount Hagen Kagam...|Papua New Guinea|HGU|
|          LAE|          AYNZ|        4|     Nadzab Airport|Papua New Guinea|LAE|
|          POM|          AYPY|        5|Port Moresby Jack...|Papua New Guinea|POM|
|          WWK|          AYWK|        6|Wewak Internation...|Papua New Guinea|WWK|
|          UAK|          BGBW|        7|   Narsarsuaq Airport|       Greenland|UAK|
|          GOH|          BGGH|        8|Godthaab / Nuuk A...|       Greenland|GOH|
|          SFJ|          BGSF|        9|Kangerlussuaq Air...|       Greenland|SFJ|
|          THU|          BGTL|       10|     Thule Air Base|       Greenland|THU|
|          AEY|          BIAR|       11|   Akureyri Airport|         Iceland|AEY|
+-------------+--------------+---------+--------------------+----------------+---+
```

This dataset contains data on flights from many countries. We can filter on just the country as USA and figure the first few lines of flights for the USA. To do this, we can use the handy filter method available on the graph dataframe:

```
gf.vertices().filter("country = 'United States'").show();
```

This will print the flights from the USA, as follows:

```
+-------------+--------------+---------+--------------------+-------------+---+
|airportIataCode|airportIcaoCode|airportId|         airportName|      country| id|
+-------------+--------------+---------+--------------------+-------------+---+
|          BTI|          PABA|     3411|Barter Island LRR...|United States|BTI|
|          LUR|          PALU|     3413|Cape Lisburne LRR...|United States|LUR|
|          PIZ|          PPIZ|     3414|Point Lay LRRS Ai...|United States|PIZ|
|          ITO|          PHTO|     3415|Hilo Internationa...|United States|ITO|
|          ORL|          KORL|     3416|Orlando Executive...|United States|ORL|
|          BTT|          PABT|     3417|    Bettles Airport|United States|BTT|
|          Z84|          PACL|     3418|      Clear Airport|United States|Z84|
|          UTO|          PAIM|     3419|Indian Mountain L...|United States|UTO|
|          FYU|          PFYU|     3420|  Fort Yukon Airport|United States|FYU|
+-------------+--------------+---------+--------------------+-------------+---+
```

Now, let's find the total number of airports in the USA. For this, we will just invoke the count method on top of the filter and this will return the count of rows, which is nothing but the count of the airports only in the USA:

```
System.out.println("Airports in India ---->" + gf.vertices().
filter("country =
'United States'").count());
```

This will print the number of airports in the USA as follows:

```
Airports in USA ----> 1435
```

We have seen how easy it is to look at the data of vertices and edges in a graph using GraphFrames. Let's dig a little deeper and find the number of flights that are going out of Newark Airport in New Jersey, which has the IATA code `'EWR'` (if you recall the number of channels going out of an entity in a graph represents the outdegrees, so, in this case, we have to find the out degrees of the airport vertice Newark (with the code `'EWR'`):

```
gf.outDegrees().filter("id = 'EWR'").show();
```

This will print the number of edges leading out of the airport `'EWR'` (or outdegrees), and this is nothing but the flights going out of this airport:

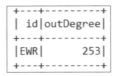

So, **253** flights are going out of the airport **EWR** according to this dataset. You can cross-verify the value you obtain by getting all the edges and checking the `src` (source) entry on all the edges; this will also give the flights that originate at the 'EWR' airport. The code for this is also a one-liner as shown next:

```
System.out.println("Flights leaving EWR airport : " + gf.edges().
filter("src =
'EWR'").count());
```

Now, let's find the top airports with the highest inbound and outbound flights count. To do this, we will first find the degree (that is total degree) of all the flights using GraphFrames. We will then register this dataset as a temporary view. Finally, we will query this temporary view, join it with the temporary view of airports and figure out the states which have the airports with the highest inbound and outbound flights count:

```
gf.degrees().createOrReplaceTempView("degrees");
session.sql("select a.airportName, a.State, a.Country, d.degree from airports a,
    degrees d where a.airportIataCode = d.id order by d.degree desc").show(10);
```

As seen here, we first pull the degrees from the GraphFrame instance, register this as a temporary view, and later fire the query to pull the data for the top 10 airports, containing the airport name, state, country, and the degree count. This will print the result as follows:

```
+--------------------+------------------+---------------+------+
|         airportName|             State|        Country|degree|
+--------------------+------------------+---------------+------+
|Hartsfield Jackso...|           Atlanta|  United States|  1826|
|Chicago O'Hare In...|           Chicago|  United States|  1108|
|Beijing Capital I...|           Beijing|          China|  1069|
|London Heathrow A...|            London| United Kingdom|  1051|
|Charles de Gaulle...|             Paris|         France|  1041|
|Frankfurt am Main...|         Frankfurt|        Germany|   990|
|Los Angeles Inter...|       Los Angeles|  United States|   990|
|Dallas Fort Worth...| Dallas-Fort Worth|  United States|   936|
|John F Kennedy In...|          New York|  United States|   911|
|Amsterdam Airport...|         Amsterdam|    Netherlands|   903|
+--------------------+------------------+---------------+------+
```

As you can see, Atlanta and Chicago have the airports with the highest number of flights. This information can actually be validated using a Google search too to check out how good our dataset and its results are. Atlanta airport is supposedly a very busy airport.

The tabular data is good, but once this data is plotted on a graph, it becomes even more comprehensible. So, let's plot the top airports on a bar chart. To do this, we will use the same code as we used earlier to plot bar charts; just the source of data and dataset creation operation will change. To maintain the brevity of the code, we are not show the code for chart creation here, but readers can find that code in the GitHub directory of the code bundle for charts:

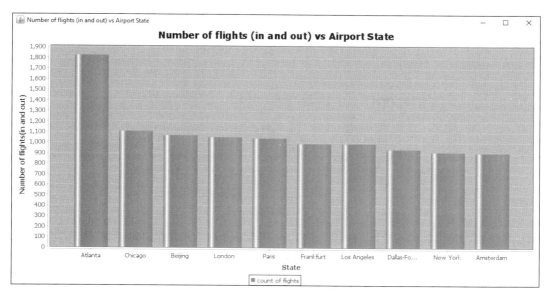

As you can see, the airport in **Atlanta** has the highest number of flights, followed by **Chicago** and **Beijing**.

Now, we will dive into a very simple but important concept of a graph, and that is the use of triplets. Three entities the source vertice, the destination vertice, and the edge combined together form a triplet. In a graph, there can be many triplets and GraphFrame provides a handy method to pull all the triplets present in a graph. Apart from giving details of the source, destination, and the edge, the triplets dataset also provides information on their respective attributes as well.

Using this triplet information, we will now find direct flights between USA and India and look at their details. To do this, we will first figure out all the triplets from the GraphFrames datasets. Next, we will filter out the triplets that have a starting vertice of country `'United States'` and a destination vertice with the country as `'India'`. We will register this dataset of triplets as a temporary view called `"US_TO_INDIA"`:

```
gf.triplets().filter("src.country = 'United States' and dst.country = 'India'")
.createOrReplaceTempView("US_TO_INDIA");
```

Next, we will query this temporary view and pull out the information for the starting city of the direct flight, the destination city, and the airline name. To do this, we will also join with the airlines temporary view to pull the airline name out from it:

```
session.sql("select u.src.statesource_city, u.dst.state destination_city,a.airlineName from US_TO_INDIA u, airlines a where u.edge.airLineCode = a.airlineId").show(50);
```

This will print out the result of the query on the console as shown next:

```
+-----------+----------------+----------------+
|source_city|destination_city|     airlineName|
+-----------+----------------+----------------+
|     Newark|          Mumbai|Air India Limited|
|    Chicago|           Delhi|Air India Limited|
|   New York|           Delhi|Air India Limited|
|     Newark|          Mumbai|  United Airlines|
|     Newark|           Delhi|  United Airlines|
+-----------+----------------+----------------+
```

As you can see from the data, there are five direct flights from the US to India. This information is actually quite accurate, as we confirmed it with data available on the internet, so this says a lot about our dataset—that it is, in fact quite good.

Let's now use a graph algorithm and see how it can help us with our trips. Earlier, we pulled a dataset of triplets. Using that dataset, we will figure out whether there is a triplet that exists between **San Francisco (SFO)** and **Buffalo (BUF)**. In other words, we are trying to figure out whether there is a direct flight between San Fransisco and Buffalo in this dataset. So, we first pull all the triplets from the GraphFrame instance, and later ,we apply the filter to figure out whether a triplet exists between the source 'SFO' and the destination 'BUF':

```
gf.triplets().filter("src.airportIataCode='SFO' and
dst.airportIataCode='BUF'").show();
```

This will print the result as follows:

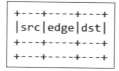

As you can see, the results are empty. Thus, there is no direct flight between SFO and BUF (Buffalo) in this dataset. If a direct flight is not available, then let's try to find a flight that takes a stop from SFO to some other airport and finally goes on to Buffalo.

To find a one-stop flight, we will use our standard graph search algorithm technique, and that is breadth first search. Using the breadth first search, we will find the vertices from SFO that are located two steps away from SFO and figure out whether one of them is 'BUF' (Buffalo). If it is, then we have found our one-stop flight to Buffalo from San Francisco. To code this, we will invoke the breadth first method on the GraphFrame instance, provide the starting vertex of the breadth first search, and the end point or end vertex of the breadth first search. We will also provide the number of levels to which the breadth first search should proceed, and this is nothing but the path. In our case, we set this value to two, as we want the search to stop at the second level (one-stop flight). Finally, after setting the parameters on the breadth first search on the graph, we will invoke the run method to run this operation on the underlying dataset.

```
Dataset<Row>sfoToBufDS = gf.bfs().fromExpr("id = 'SFO'").toExpr("id =
'BUF'").maxPathLength(2).run();
```

As you can see, the result of the breadth first search is also in the form of a dataset and we store it in a variable. Next, we register this dataset as a temporary view called sfo_to_buf:

```
sfoToBufDS.createOrReplaceTempView("sfo_to_buf");
```

Finally, we will query on top of this temporary view to figure out the starting state of the vertice, the connecting state, and finally, the state where the flight ends (in our case, this is Buffalo). We will also print the output to the console by invoking the show method.

```
session.sql("select distinct from.state , v1.state, to.state from
sfo_to_buf").show(100);
```

This will print the result on the screen as follows:

```
+-------------+---------------+-------+
|        state|          state|  state|
+-------------+---------------+-------+
|San Francisco|    Minneapolis|Buffalo|
|San Francisco|      Charlotte|Buffalo|
|San Francisco|        Orlando|Buffalo|
|San Francisco|      Las Vegas|Buffalo|
|San Francisco|         Newark|Buffalo|
|San Francisco|   Philadelphia|Buffalo|
|San Francisco|        Atlanta|Buffalo|
|San Francisco|        Phoenix|Buffalo|
|San Francisco|     Washington|Buffalo|
|San Francisco|        Detroit|Buffalo|
|San Francisco|         Boston|Buffalo|
|San Francisco|      Baltimore|Buffalo|
|San Francisco|        Chicago|Buffalo|
|San Francisco|Fort Lauderdale|Buffalo|
|San Francisco|       New York|Buffalo|
|San Francisco|      Cleveland|Buffalo|
+-------------+---------------+-------+
```

As you can see, even though there is no direct flight to **Buffalo** from **San Francisco**, still, you can take many single-hop flights, depending upon your requirements. For example, you can take a flight to **Washington** and from there take another flight to **Buffalo**.

Next, we will find the most important airports present in this dataset. To calculate the importance of an airport, we will use the PageRank algorithm. As you will recall, this algorithm is the same one that is used by Google for their search engine. It is used to give importance to a website link and, based on that importance, links show up on the search page. To calculate importance, the algorithm checks for the connections of a website to other websites and checks how important the other websites are. The concept is that an important website will link to another important website only. PageRank can be used in other cases as well as in a graph to figure out the importance of vertices, based on how they are connected to other vertices.

The best part about the usage of this algorithm is that it is bundled by default in the GraphFrame library and can be used directly by invoking it on the GraphFrame instance. So, we invoke the PageRank method on the GraphFrame instance, supply the necessary parameters of the maximum number of iterations (we keep this at five; you can play with this value), and reset the probability as 0.15. The code to run PageRank on the graph is a single-liner as shown next:

```
Dataset pg = gf.pageRank().resetProbability(0.15).maxIter(5).run().
vertices();
```

As you can see, we invoked the PageRank method on GraphFrame, then supplied the necessary parameters, and invoked the run method to run the algorithm on the underlying dataset. Finally, we extract the vertices from the result obtained from the PageRank algorithm. This result is shown next:

```
+---------+--------------------+-------------+---+------------------+-------------------+----------------+-------------------+
|airportId|         airportName|      country| id|          latitude|          longitude|           state|           pagerank|
+---------+--------------------+-------------+---+------------------+-------------------+----------------+-------------------+
|     2989|    Syktyvkar Airport|       Russia|SCW|  61.64699935913086|   50.84510040283203|       Syktyvkar| 0.41237783214480783|
|     1686|    ?ncirlik Air Base|       Turkey|UAB|  37.002101898199996|       35.4258995056|           Adana|                0.15|
|     2548|    Marechal Rondon A...|    Brazil|CGB|      -15.6528997421|      -56.1166992188|          Cuiaba|  1.5706549071102351|
|     8284| Capital City Airport|United States|CXY|      40.2170982361|   -76.85150146480001|      Harrisburg|                0.15|
|     3368|    Tianjin Binhai In...|     China|TSN|  39.124401092499994|         117.346000671|         Tianjin|  1.0415126053412012|
|     1837|    Puerto Escondido ...|    Mexico|PXM|      15.8768997192|  -97.08910369870001|Puerto Escondido| 0.19125829670841465|
|     8793|        Hardwick Field|United States|HDI|  35.22010040283203|     -84.8323974609375|       Cleveland|                0.15|
|      868|     Virginia Airport| South Africa|VIR| -29.770599365234375|   31.058399200439453|          Durban|                0.15|
|     6917|    Antonio Nery Juar...| Puerto Rico|ARE|      18.4500007629|     -66.6753005981|         Arecibo|                0.15|
|     6013|         Jolo Airport|  Philippines|JOL|   6.0536699295043945|   121.01100158691406|            Jolo| 0.20276920684098537|
|      128|  Resolute Bay Airport|       Canada|YRB|       74.7169036865|       -94.9693984985|        Resolute|  0.5402124513888888|
|      413|    Kuressaare Airport|      Estonia|URE|   58.22990036010742|    22.50950050354004|      Kuressaare| 0.16411268502907458|
+---------+--------------------+-------------+---+------------------+-------------------+----------------+-------------------+
```

As you can see, the last column has the PageRank calculated value. The greater the value, the better the PageRank of the entity is. The values are distributed as per the rows in the dataset. We need a better way to query the results and figure out the top entries of the airports with the highest PageRank. To do this, we will register the resulting dataset as a temporary view and fire a query to pull out all the details in the descending order of their PageRank, where the highest ranked are listed first and then the remaining ones.

```
pg.createOrReplaceTempView("pageranks");

session.sql("select * from pageranks order by pagerankdesc").show(20);
```

[315]

This will print the result as follows:

```
+---------------+---------------+---------+--------------------+------------------+----+----------------+-------------------+
|airportIataCode|airportIcaoCode|airportId|         airportName|           country|  id|           state|           pagerank|
+---------------+---------------+---------+--------------------+------------------+----+----------------+-------------------+
|            ATL|           KATL|     3682|Hartsfield Jackso...|     United States| ATL|         Atlanta| 48.866507796851174|
|            ORD|           KORD|     3830|Chicago O'Hare In...|     United States| ORD|         Chicago| 30.85875567259695|
|            LAX|           KLAX|     3484|Los Angeles Inter...|     United States| LAX|     Los Angeles| 29.767643411164528|
|            DFW|           KDFW|     3670|Dallas Fort Worth...|     United States| DFW|Dallas-Fort Worth| 28.614101408262563|
|            SIN|           WSSS|     3316|Singapore Changi ...|         Singapore| SIN|       Singapore| 25.836970924753565|
|            CDG|           LFPG|     1382|Charles de Gaulle...|            France| CDG|           Paris| 25.494167226690674|
|            LHR|           EGLL|      507|London Heathrow A...|    United Kingdom| LHR|          London| 25.258003242524943|
|            DEN|           KDEN|     3751|Denver Internatio...|     United States| DEN|          Denver| 24.906201842614077|
|            DME|           UUDD|     4029|Domodedovo Intern...|            Russia| DME|          Moscow| 23.29190755405923|
|            JFK|           KJFK|     3797|John F Kennedy In...|     United States| JFK|        New York| 23.064144993605314|
|            FRA|           EDDF|      340|Frankfurt am Main...|           Germany| FRA|       Frankfurt| 22.714716381995252|
|            PEK|           ZBAA|     3364|Beijing Capital I...|             China| PEK|         Beijing| 22.654135319752914|
|            SYD|           YSSY|     3361|Sydney Kingsford ...|         Australia| SYD|          Sydney| 22.603944128539965|
|            MIA|           KMIA|     3576|Miami Internation...|     United States| MIA|           Miami| 21.973498718530653|
|            AMS|           EHAM|      580|Amsterdam Airport...|       Netherlands| AMS|       Amsterdam| 21.067548191034817|
|            IST|           LTBA|     1701|Atatürk Internati...|            Turkey| IST|        Istanbul| 20.923165962750826|
|            DXB|           OMDB|     2188|Dubai Internation...|United Arab Emirates| DXB|           Dubai| 20.764004952251774|
|            BKK|           VTBS|     3885|Suvarnabhumi Airport|          Thailand| BKK|         Bangkok| 19.537859448118365|
|            BOG|           SKBO|     2709|El Dorado Interna...|          Colombia| BOG|          Bogota| 19.471296723900714|
|            ICN|           RKSI|     3930|Incheon Internati...|       South Korea| ICN|           Seoul| 18.865901630530471|
+---------------+---------------+---------+--------------------+------------------+----+----------------+-------------------+
only showing top 20 rows
```

As you can see, the most important airports are Atlanta airport, followed by Chicago, and so on.

The graphs that we have dealt with in this dataset are huge and have lots of edges. If you will check a graph on a social network, it will be much, much bigger. Oftentimes, the need is to analyze a smaller dataset. GraphFrame provides a functionality out-of-the-box, by means of which we can break the gigantic graphs into smaller manageable graphs so that we can analyze them separately. This approach is known as pulling out subgraphs.

The concept is very simple from the point of view of the GraphFrames library. As the GraphFrame instance is built using a dataset of vertices and a dataset of edges, we can always build a fresh instance of GraphFrames by providing a fresh instance of vertices and edges. For the purpose of a subgraph, we can filter out the big vertices and extract the vertices we are interested in; and, similarly, we can filter out the edges we need and build a GraphFrame object with those.

In the code example shown next, we will filter out the vertices of United Kingdom, that is, we are only extracting the airports of United Kingdom. We also register the vertices dataset as a temporary view so that we can further query on it:

```
Dataset<Row>ukVertices = gf.vertices().filter("country = 'United
Kingdom'");
    ukVertices.createOrReplaceTempView("ukAirports");
```

Chapter 11

We also pull the edges dataset from the original GraphFrame object and register it as a temporary view as well:

```
gf.edges().createOrReplaceTempView("ALL_EDGES");
```

Next, we filter the edges that do not start and end in the United Kingdom as we are only interested in the airports in the United Kingdom, and the domestic flights that flow within the United Kingdom:

```
Dataset<Row>ukEdges = session.sql("select * from ALL_EDGES where srcId
in (select airportId from ukAirports) and dstId in (select airportId
from ukAirports)");
```

We can now build our `GraphFrame` object by creating the new GraphFrame instance with these vertices and edges specific to airports in the United Kingdom only:

```
GraphFrameukSubGraph = newGraphFrame(filtered, ukEdges);
```

We can selectively use this subgraph to run selective analytics only on the United Kingdom data.

Let's now plot this graph using the GraphStream library to see how it looks on the user interface. To do this, we will create an instance of the GraphStream library:

```
Graph graph = newSingleGraph("SimpleGraph");
```

We will collect the vertices of the United Kingdom subgraph, iterate over them and populate the graph nodes of the GraphStream instance. We keep the airport state location as the name of the node:

```
List<Row>ukv = ukSubGraph.vertices().collectAsList();
for (Row row : ukv) {
graph.addNode(row.getString(2)).setAttribute("ui.label", row.
getString(8));
}
```

We will also collect the edges data from the subgraph, iterate over this list, and populate the edges of the graph with the airport ID from where the edge starts and the airport ID from where the edge ends.

```
List<Row>uke = ukSubGraph.edges().distinct().collectAsList();
Map<String, String>keys = newjava.util.HashMap<>();
for (Row row : uke) {
if(!keys.containsKey(row.getString(9) + row.getString(5))) {
graph.addEdge(row.getString(9) + row.getString(5), row.getString(9),
row.getString(5));
   keys.put(row.getString(9) + row.getString(5), "KEY");
}
}
```

This would populate the graph of GraphStream with proper data for nodes and edges.

> As you might have seen in the preceding code, we have used a HashMap to filter out duplicate edges from being added to the graph edges. This is done because of an error that was coming from GraphStream due to duplicate keys.

Finally, you can display this graph on the screen using the display method of GraphStream:

```
graph.display();
```

This would print the graph on the display as follows:

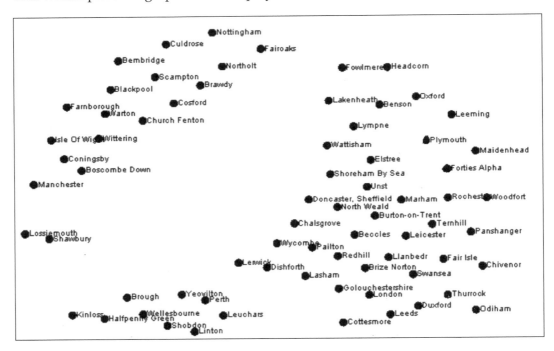

As you can see, all the nodes belong to United Kingdom only. There are some empty edges because the routes dataset that we have do not have any connecting flights between these airports.

From the image you can notice there are clearly connected nodes that are meshed up a bit. We're leaving it to you to further beautify this graph by filtering out the UK airports that do not appear in the routes at all and only use the airports in the vertices that appear in the edges, as this would print a better graph for display.

With this we come to an end of our chapter on building graphs on massive datasets. Graph analytics is a big topic and certainly cannot be covered in just one chapter of a book—an entire book can be written on this topic. Here are some of the resources that we would recommend if you want to explore further in the field of graph analytics:

- Learning Neo4j book from Packt publishers
- Free course on Graph Algorithms from Stanford in Coursera
- Graph analytics on big data course on Coursera
- Apache Spark graph processing book from Packt publishers
- Mining of massive datasets free book by Stanford

Summary

In this chapter, we learned about graph analytics. We saw how graphs can be built even on top of massive big datasets. We learned how Apache Spark can be used to build these massive graphs and in the process we learned about the new library GraphFrames that helps us in building these graphs. We started with the basics of graphs as to how graphs can be built and represented and later we explored the different forms of analytics that can be run on those graphs be it path-based analytics involving algorithms such as breadth first search or connectivity analytics involving the degrees of connection. A flight dataset was used to explore the different forms of graph analytics while using a real-world dataset.

Up until now, we have mostly used the data and the program in a batch mode. In the next chapter, we will see how big data can even be used in our analysis at real time.

12
Real-Time Analytics on Big Data

At some point in time we might all have used insurance quotes. To get insurance quotes for a car we fill in the details about us and based on our credit history and other details the application gives you the insurance quotes in real time. This application analyzes your data in real time and based on it predicts the quotes. For years, these applications have followed mostly rule-based approaches with a powerful rule engine running behind the scenes, more recently these applications have started using machine learning to analyze data further and make predictions at that point in time. All these predictions and analysis that happen at that instance or point in time are real-time analytics. Some of the most popular websites, such as Netflix or famous ad networks, are all using real-time analytics and with the coming of new devices as part of the Internet of things or IoT wave, collection and analysis of data in real time has become the need of the hour. Now, we need powerful products that can tell us how the air quality in a zone at that instance in time is or the traffic status on a particular highway at a particular time.

With real-time analytics in mind in this chapter, we will cover the following:

- Basic concepts of real-time analytics and covering some real-world use cases.
- How business analysts or data scientists can run fast performing queries in real time on big data. Here we will cover brief introduction on products such as Apache Impala.
- Basic concept of Apache Kafka and how it gels between data collectors and data analysers.
- Spark Streaming and how it can be used to analyze the data from Apache Kafka.

- Finally, we will cover some sample case studies as follows:
    - Trending videos in real time
    - Sentimental analyses on tweets in real time
- Before we get into the details of real-time analytics, let's dive into some of the main concepts of real-time analytics.

Real-time analytics

As is evident from the name, real-time analytics provides analysis and their results in real time. Big data has mostly been used in batch mode where the queries on top of the data run for a long time and the result is later analysed. The approach is changing lately, mainly due to the new requirements pertaining to certain use cases that require immediate results. Real-time requires a separate set of architecture that caters to not only data collection and data parsing, but also data analyzing at the same time.

Let's try to understand the concept of real-time analytics using the following diagram:

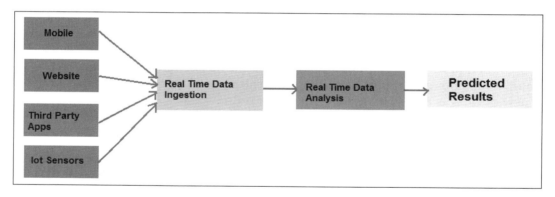

As you can see, today the sources of data are plenty whether it's mobile devices, websites, third-party applications, or even the Internet of Things (sensors). All this data needs a way to propagate and flow from the source of their devices to the central unit where the data can be parsed, cleaned, and finally ingested. It is at this ingestion time that the data can also be analyzed and deductions can be made from it. The analysis made on the real-time data can be of various kinds whether predictive analysis, or general statistics such as figuring out quantiles or aggregations over a period of time or within a time window.

There are certain clear use cases of big data real-time analytics due to which this type of analytical approach has picked up so much steam. Some of those use cases are:

- **Fraud analytics**: Detecting fraud in real time is a high priority and one of the top use cases of real-time analytics. Fraud that is detected a few days late might already have done the damage, therefore, catching it earlier on and taking the appropriate action might save data loss or loss of any other resources, such as money. One of the simplest examples could be catching credit card fraud in real time.
- **Sensor data analysis (Internet of Things)**: With the usage of Internet of Things on the rise there are a lot of devices such as sensors, and so on, which not only collect data in real time, but also transmit it. This data can be picked up and analyzed in real time. For example, in the case of smart cities this data from sensors can tell the amount of air pollutants in the air at a certain point in time.
- **Giving recommendations to users in real time**: By analyzing the click stream of the users in real time the current interests of the users can be figured out and accordingly recommendations can be made to the users of a website or mobile app in real time
- **Real-time analytics in healthcare**: We have so many healthcare devices that monitor healthcare statistics in real time, for example, the wearables that we wear, pacemakers installed inside bodies, and so on. All these devices emit data that can be tracked, ingested, and monitored. This is a huge amount of data that needs to be tracked in real time to generate alerts, for example in the case of abnormal conditions.
- **Ad-processing in real time**: Internet ads can be processed in real time using the big data stack. The system can check the user's historical patterns of net access and based on that show exclusive ads targeting the customers.

Apart from the use cases shown in the preceding list there are plenty of other use cases that require real-time analytics usage.

Real-time analytics require a different set of products from the big data stack and we will introduce some of those in the next section.

Big data stack for real-time analytics

For real time data ingestion, processing, and analysis we need a separate set of products from the big data stack. The big data stack is very large and whole lot of private as well as open source vendors are involved. As such, there are plenty of products in this space. Covering all of these products is beyond scope of this book, but we will give a very brief introduction to products and solutions that we will be using in this chapter.

There are various products suitable for each purpose whether real time SQL queries, real time data ingestion, or parsing streaming data. Some of the popular ones are:

- Apache Impala
- Apache Drill
- Apache Kafka
- Apache Flume
- HBase
- Cassandra
- Apache Spark (Streaming Module)
- Apache Storm

There are many more products apart from the ones mentioned previously. Each of the product works in a different category of its own and we will be giving a brief introduction to these products in the upcoming sections.

Real-time SQL queries on big data

We have mostly used Spark-SQL in this book until now and we also mentioned that Apache Hive can be used to fire SQL queries on big data. But mostly all these queries and products are batch only. That is to say they are slow to use on big data. We need a separate set of products if we need to fire fast SQL queries on big data and those that have performance similar to SQL query tools on RDBMS. There are now tools available that help us fire fast SQL queries on top of your big data. Some of the popular ones are:

- **Impala**: This product is built by the company Cloudera and it helps in firing real time fast queries on big data. We will cover this briefly in our next section.
- **Apache Drill**: Apache Drill is another product that is used to fire real time SQL queries on top of big data. The underlying data can lie in any of the different systems and formats for example the data can reside in HBase, HDFS, Cassandra, or Parquet. Apache Apache Drill has a unique architecture that helps it in firing fast SQL queries on top of this data.

Real-time data ingestion and storage

There are also many products for data ingestion. These products connect to the source of data and help in pulling data from them and transferring to the big data systems like HDFS or NoSQL databases like HBase and Cassandra. Some of these products are:

- **Apache Kafka**: Apache Kafka is a messaging technology similar to IBM MQSeries or RabbitMQ except that it runs on big data and is highly scalable due to its architecture. We will cover Kafka briefly in one of our upcoming sections in this chapter.
- **Apache Flume**: Flume is a distributed and reliable service that can pick huge streaming data from various sources and push to HDFS. We will not be covering flume in this book.
- **HBase**: This is a massively scalable NoSQL database that runs on top of HDFS and data that is gathered from sources such as Kafka and flume can be directly pushed to it for analysis
- **Cassandra**: Flume is a distributed and reliable service that can pick huge streaming data from various sources and push to HDFS. We will not be covering Flume in this book.

Real-time data processing

There are products in the market that read the data from the stream as it arrives. Thereby these process the data in real time. They might clean the data at that point in time and ingest it into NoSQL databases such as HBase or Cassandra. There are a few products in this space, but two stand out on the big data stack and are quite popular:

- **Spark Streaming**: Spark has a module dedicated to dealing with real-time data analysis. This module can be integrated with real-time data and can pull in data from streams and analyze them. We will cover this in detail in our next section.
- **Storm**: This is another product for real-time data analysis.

A typical example of a real-time processing system is shown in the following figure:

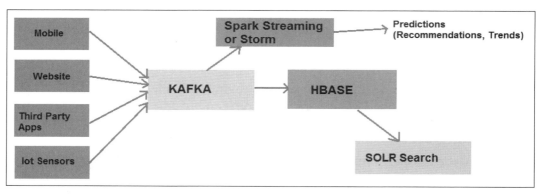

The preceding architecture diagram shows that real-world architecture can be much more difficult than this. According to this architecture diagram, there are various sources of data such as mobile phones, third-party apps, and IoT sensors. All of these can push their data to Kafka topics and data is thus transferred from the sources. Using Spark Streaming we can read this data in mini batches from the Kafka topics. This data can be cleaned and analyzed and at this time predictive models that were pre-trained can also be applied on this data for predictive results. The data can also be stored into a NoSQL database such as **HBase** for further analysis. The data in **HBase** can be further indexed and used by a search framework such as **SOLR** for firing search queries.

In the next section, let's use one of the SQL query engines on Hadoop and use it for some simple data analysis in real time.

Real-time SQL queries using Impala

Impala is a very high performing SQL query engine that runs on top of big data that is stored in **Hadoop Distributed File System** (**HDFS**). This tool is frequently used by data scientists and business analysts who want to quickly query their big data and do their analysis or generate reports. The other SQL query engines on top of big data execute MapReduce jobs when SQL queries are fired, but Impala does not follow this approach; instead it relies on its daemon process to cache data as well as to process the queries. It is therefore much faster than traditional products such as Hive. Another similar product is Apache Drill, which also provides faster SQL query performance on top of big data:

There are a few key advantages of using Impala:

- It has a very high performance on big data and its performance becomes better if the underlying data is in compressed and columnar format like in Parquet format.
- It can pull data from various data types on HDFS. For example, we can create tables in Impala for data in Parquet format in HDFS, or we can create tables in Impala for data in JSON format. It can even pull data directly from HBase.
- It provides JDBC support on Java and thereby using the Java API we can build Java applications that directly fire queries on Impala tables.
- It can be integrated with Apache kudu for fast analytics on big data. Apache Kudu is another product which is HDFS-like and is optimized specifically for fast analytics using Impala.

While covering Impala in detail is beyond the scope of this book, we will quickly cover a simple case study to show how powerful Impala is.

Flight delay analysis using Impala

This is a simple application that we will use in our analysis for firing SQL queries in real time. We will be using the flight delay dataset (available openly from The Bureau of Transportation statistics on this link https://www.transtats.bts.gov/DL_SelectFields.asp?Table_ID=236&DB_Short_Name=On-Time).This dataset has a lot of columns, the columns that we will be using in our analysis are:

Attribute	Description
AirlineID	An identification number for a unique airline carrier
Origin	Origin airport
OriginState	Origin airport state code
DepDelay	Difference in minutes between scheduled and actual departure time
Dest	Destination airport
DestState	Destination airport state code
ArrDelay	Difference in minutes between schedule and actual arrival time
Distance	Distance between airports (miles)
Weather_delay	Weather delay in minutes
Cancelled	Cancelled flight indicator (1 = Yes)

We first need to load this data in HDFS. You can directly copy the file to HDFS, but our aim is to expedite our SQL queries on top of this data. Due to this we first convert our data to Parquet and then we store it in HDFS. We converted the data to Parquet first because Parquet stores data in columnar format and if we need to run queries only on a few subset of columns it is way faster than the other formats like plain text, CSV or even row formats like Avro. Also since data is stored in columnar format so if an aggregation needs to be run on a few columns it is very fast as data in a single column will be of the same type in Parquet and can be clubbed and aggregated together easily.

Converting the data to Parquet can be done through a simple Spark Java program. We will build the Spark context and load the data of the target dataset into a Spark `dataset` object. Next we register this Spark dataset as a temporary view. Finally, we select the attributes from the dataset we are interested in (we also provide good names for the columns) using an SQL query. The results of this SQL query are again stored in a dataset. We push the contents of this dataset into an external Parquet file:

```
Dataset<Row>rawDataAirline =
session.read().csv("data/flight/On_Time_On_Time_Performance_2017_1.
csv");
rawDataAirline.createOrReplaceTempView("all_data");

Dataset<Row>filteredData = session.sql("select _c8 airline_code, _c14
src, _c16
src_state, _c31 src_delay, c23 dst, _c25 dst_state, _c42 dst_delay,_
c54
distance, _c57 weather_delay, _c47 cancelled from all_data");

filteredData.write().parquet("<HDFS_FILE_PATH>");
```

Once the data is stored in HDFS, an external table can be created by Impala and pointing to the source of this Parquet file location in HDFS. Creating an external table in Impala is simple and can be done using the command shown as follows:

```
create external table logs (field1 string, field2 string, field3 string)
partitioned by (year string, month string, day string, host string)
row format delimited fields terminated by ','
location '/user/impala/data/logs';
```

There are a few important concepts depicted in the SQL statement for creating the preceding external table. We will briefly touch on those in the following table:

SQL key word	Description
external	The `external` keyword in the create table statement signifies that this is an external table. That is to say the source of data of this table lies outside like in HDFS. So if you manually delete a folder of this data from HDFS then the data for the table is also deleted (you do need to refresh the external table once for this).
Column name	We need to provide the names of the columns and data types. For detailed info on the data type supported refer to Impala documentation.
partitioned	This is a very important concept that helps in scalability and performance of content on top of HDFS. In simple terms it means cutting the data into chunks and storing it in separate folders with particular names. Queries can then be made to be specific based on certain data types that will look for particular folders only while leaving some other folders. Therefore, the queries will become faster as they now only have to touch a subset of data instead of an entire set of dataset.
location	This is location on HDFS where the actual external file or folder containing the files is present on HDFS. Its is important to note that file type can be of any type like JSON, text or even Parquet.

After setting up the external table it is time to query it using the Impala query engine. Impala provides a JDBC interface using which you can access the data in the table. For this you can write your Java program accessing the data from the Impala table. You can also execute the SQL queries directly from the `hue` tool provided by Cloudera and this tool can be used to fire queries on top of the Impala tables. The Java program for database access on Impala has regular boiler plate code and can be checked on the Cloudera website.

We can fire simple analytical queries to check how speedy the Impala query engine is and whether it is good to use for your use cases in real time.

> Hue is a web application that comes deployed with the Cloudera bundle (it is an open source tool) and it gives an interface to your underlying data systems like Impala and HBase. You can fire queries to these data systems using Hue.

Real-Time Analytics on Big Data

Let's fire some analytical queries on our Impala table. First let's check how many unique flights we have in this dataset. For this we will count the distinct source and destination combinations in this dataset using the following query.

```
select count(*) from (select distinct origin,dest from flights)
```

This will print the total unique flights in our `flights` table.

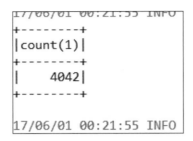

You can also verify the speed of the Impala query, it should not take more than one or two seconds. Let's now find a first few flights that got delayed by more than 30 minutes from the SFO airport.

```
select * from flights where dep_delay > 30 and origin = 'SFO'
```

This will print the result as:

```
17/06/01 00:55:45 INFO DAGScheduler: Job 3 finished: show at CsvToParquet.java:32, took 0.471550 s
+----------+------+------------+---------+----+----------+---------+---------+-------------+---------+
|airline_id|origin|origin_state|dep_delay|dest|dest_state|arr_delay|distance |weather_delay|cancelled|
+----------+------+------------+---------+----+----------+---------+---------+-------------+---------+
|     19805|   SFO|          CA|    53.00| PHX|        AZ|    50.00|   651.00|         0.00|     0.00|
|     19805|   SFO|          CA|    38.00| PHX|        AZ|    24.00|   651.00|         0.00|     0.00|
|     19805|   SFO|          CA|   142.00| PHX|        AZ|   156.00|   651.00|         2.00|     0.00|
|     19805|   SFO|          CA|    52.00| DFW|        TX|    46.00|  1464.00|         0.00|     0.00|
|     19805|   SFO|          CA|   100.00| DFW|        TX|   103.00|  1464.00|         0.00|     0.00|
|     19805|   SFO|          CA|    46.00| DFW|        TX|    31.00|  1464.00|         0.00|     0.00|
|     19805|   SFO|          CA|    85.00| DFW|        TX|    99.00|  1464.00|         0.00|     0.00|
|     19805|   SFO|          CA|    31.00| DFW|        TX|    18.00|  1464.00|         0.00|     0.00|
|     19805|   SFO|          CA|   294.00| PHX|        AZ|   291.00|   651.00|         0.00|     0.00|
|     19805|   SFO|          CA|    46.00| ORD|        IL|     8.00|  1846.00|         null|     0.00|
|     19805|   SFO|          CA|    54.00| ORD|        IL|    53.00|  1846.00|         0.00|     0.00|
|     19805|   SFO|          CA|    56.00| MIA|        FL|    32.00|  2585.00|         0.00|     0.00|
|     19805|   SFO|          CA|    33.00| MIA|        FL|     7.00|  2585.00|         null|     0.00|
|     19805|   SFO|          CA|    49.00| MIA|        FL|    58.00|  2585.00|         0.00|     0.00|
```

As you can see previously a lot of flights did get delayed by more than thirty minutes and some of them were flying to Phoenix and Orlando.

Finally, let's run a group by query and count the number of flights cancelled in each state and order it in descending order. So this way we have the state with maximum flights cancelled listed first followed by other states and their flight cancelled count in that order.

```
select origin_state, count(*) cnt from flights where cancelled > 0
group by
origin_state order by cnt desc
```

This will print the result as:

```
17/06/01 01:10:11 IN
+------------+----+
|origin_state| cnt|
+------------+----+
|          CA|1425|
|          FL| 709|
|          GA| 676|
|          TX| 617|
|          IL| 554|
|          CO| 458|
|          NY| 438|
|          MI| 342|
|          NC| 311|
|          OR| 272|
```

As you can see previously California and Florida had the maximum number of cancelled flights as per this dataset. With this we conclude our section on discovering Impala. A bigger coverage on Impala is beyond scope of this book and we would recommend that users check out Impala and similar products like Apache Drill. They have improved a lot over the years and can fire very fast queries on top of big data now.

In the next section we will briefly introduce two technologies that are used heavily in real-time analytics nowadays and they are Apache Kafka and Spark Streaming

Apache Kafka

Kafka is a distributed high-throughput messaging system. Like other messaging systems it decouples the applications that want to interact with each other, so that they can send messages to each other using Apache Kafka. It is massively scalable as it supports partitioning across multiple nodes and unlike traditional messaging technologies like IBM MQSeries it does not store any specific state (state storage for example for message delivery confirmation is all left to the client invoking the broker). It does support replication and is fault tolerant too.

Kafka follows the publish-subscribe mechanism for data transfer. Thus, messages are pushed to a Kafka topic and there can be multiple consumers for that topic that can receive the data. We show an example of this approach in the following image:

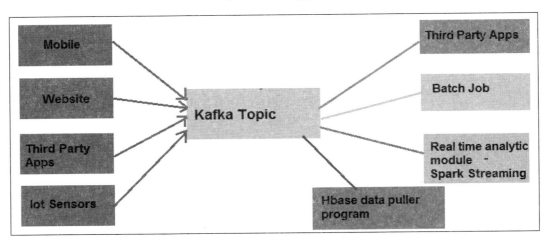

As you can see in the preceding image multiple systems have subscribed to a Kafka topic (to keep things simple we have omitted any extra module on the source of the data side). Thus, multiple systems whether they are IoT sensors, third party apps or websites can push their data to a Kafka topic. As shown in the preceding image multiple subscribers can pull data from this topic and use this data. In the image, we showed the same data can be used by third party apps as well as by real-time analytic system that might be making real time predictions on this data.

Kafka is used in real life use cases in a lot of places. Listed below are some use cases where Kafka can be used:

- **Integration with IoT sensors**: The Internet of Things is becoming very popular these days, with a lot of devices like sensors that continuously collect data from various sources. Kafka can be a mechanism used to transport this massive data in real time from various such sensors into the Hadoop stack. Data can be pulled from IoT sensors and pushed into Kafka topics.
- **Social media real-time analytics**: Social media data can also be collected and pushed into Kafka and transported for analysis. For example, tweets from Twitter can be pushed into Kafka topics based on hashtags and can be later analysed in real time.
- **Healthcare analytics**: Apart from regular healthcare apps where data be transferred across different applications using Kafka, the data collected from the newer wearable devices can also be pushed to Apache Kafka topics for further analytics.

- **Log analytics**: There is valuable information present in the log files that can help us predict things like when the application might break next or when can we expect our load balancer to go down next. Data from the logs can be streamed into Kafka topics and consumed.
- **Risk aggregation in finance**: Kafka can be used to store risk information generated in real time and can transfer it across applications in the finance field.

Next, we will see how to consume this data as it arrives in a Kafka topic in real time using another Spark module that caters to real time data.

Spark Streaming

Streaming refers to the concept of receiving data as soon as it arrives and acting on it. There are many sources that generate data in real time be they health care devices, social networks, or IoT devices and there are many mechanisms (like Apache Kafka, Flume) to transport this data as it is generated and push it to HDFS, HBase or any other big data technology. Spark streaming refers to the technology of consumed streamed data through Apache Spark, it is a separate module in Apache Spark API for real time data processing.

The concept of Spark Streaming is simple and is shown in the following image:

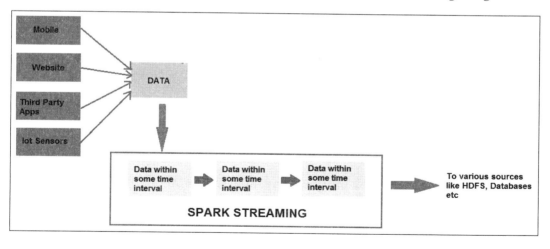

As you can see in the preceding image there might be various sources of data like mobile phones and they transfer their data as streams (using Kafka, Flume or any other technology). These streams of data are read at regular intervals as micro batches by the Spark Streaming module. Within a span of time the stream of data is represented as an RDD so Spark streaming receives a sequence of RDD. A whole lot of operations like Map, FlatMap, and filter can be done on RDDs as we already saw in chapter one, so the RDDs obtained via streaming can also be processed using these operations. Apart from these Spark streaming supports applying machine learning programs and graph algorithms to the streaming RDDs in real time.

Internally in Spark API this continuous sequence of RDDs is represented by an abstraction called as DStreams, so a DStream encapsulates this distributed dataset of RDDs and is essentially a sequence of RDDs. As seen in the preceding image since this DStream has a sequence of RDDs as such it has methods like `foreachRdd` to iterate over these RDDs. Once the RDDs are pulled from the DStream, all the regular RDD functions, such as filter, Map or FlatMap can be applied on them for further analysis.

> Even though the name says Spark streaming, Spark streaming is not true streaming as it uses micro batches as mentioned previously. '**Storm**' another popular product from Apache is truer form of streaming and is very popular. In this book we will not be covering Apache Storm, but we would recommend that the readers check out this product.

For gathering streaming data, the Spark Streaming module can be connected with various data sources like Kafka, Flume and other third party components. To integrate a component like Kafka with Spark Streaming you need to use the proper jars specific to that component and integrate Spark Streaming with it. For readers we have included a simple maven project in our GitHub repository and this would be a pre-integrated project of Spark Streaming with Kafka that the readers can use for their learning purposes.

For more detailed information on Spark Streaming please refer to the official Spark documentation. Let's now see the typical use cases where Spark Streaming can be used.

Typical uses of Spark Streaming

Spark Streaming has few typical use cases where it can be directly used. Some of them are listed as follows:

- **Data collection and storage in real time**: As mentioned earlier Spark Streaming can be connected with various sources of real time data be they social media, click stream data from websites or mobiles, or data from sensors (Internet of Things). Data has to be shipped into products like Apache Kafka, Flume, and so on and these products can then be integrated with Spark Streaming. Spark Streaming can be used to clean and transform this data and can also push it further to other databases or file systems.

- **Predictive analytics in real time**: This comes into picture when data is already streamed and available in the Spark Streaming program for analysis. Pretrained machine learning models can be applied in real time on the streamed data. For example sentiment analysis can be done on tweets from Twitter in real time.

- **Windowed calculations**: This is a very handy feature from Spark Streaming. This allows you to collect the data in a predefined time intervals (window) and do calculations on top of it. So if you want to see the clicks generated on the website in last hour you can keep a time window of an hour and Spark Streaming will conveniently split the data into the desired time window of an hour. Once the data is available the data is in the form of RDDs and these RDDs can be pulled out from the streams of data for further analysis like the most used features of the website in the last hour.

- **Cumulative calculations**: If you want to keep cumulative or running statistics in real time then out of the box Spark provides a feature for this. Spark Streaming provides a feature to store the state and the new data available can then be appended on this state for analysis. So suppose you have clickstream data where in state you have stored the number of clicks received on a feature like 'Trending Videos' then when the next micro batch of data arrives Spark Streaming can extract the new hit counts on this feature from the micro batch and can append to the stored state for a latest net hit count on this feature at that time.

So much for the theory let's get into the code now. We will now try to understand some real time use case while using some code and while doing that we will also learn about Spark Streaming and real time predictive analytics and also get information on Kafka integration with Spark.

Base project setup

We have built a sample skeleton project that we will be using for our real-time analysis use case. Our GitHub repository will contain the code pertaining to all our use cases for real time analysis within this project.

> This project setup is for readers to try out these components on a Windows OS. If you already have access to a Hadoop installation with all the components like Kafka, Spark, HDFS setup, you will probably not need this skeleton setup and can directly run your Spark Streaming code talking to these components and can also skip this section entirely.

For this base project setup we have a maven Java project with all the dependencies being part of the `pom.xml` file. The setup of our project is shown in the following image:

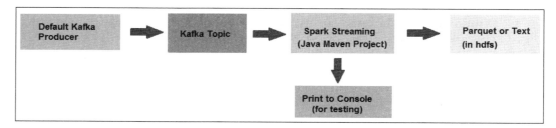

As you can see above our setup is simple. Let's go through the main points of our setup

- **We are using the default Kafka configurations**: To test our code, you can download and install the Kafka locally. Refer to the their quickstart guide and start the `zookeeper` server, `kafka-server` and finally the default Kafka producer program that ships along with the Kafka installation. While explaining each component in detail is beyond scope of this book we will briefly mention the steps for starting each component:

    1. First start the `zookeeper` server with default configuration (that is using default `zookeeper.properties` file).

       `zookeeper-server-start.bat <CONFIG_DIRECTORY>\zookeeper.properties`

    2. Next start the `kafka-server` with default Kafka configuration (that is using the default `server.properties` file)

       `kafka-server-start.bat <CONFIG_DIRECTORY>\server.properties`

3. Now create a topic on Kafka. We call our topic as `test11` and we set it up with default parameters with no replications and single partition.

   ```
   kafka-topics.bat --create --zookeeper localhost:2181
   --replication-factor 1 --partitions 1 --topic test11
   ```

4. This will be the topic where you will publish your data and read from Spark Streaming program. The data on this topic can be anything like tweets, input on a form on a website and can be in any format like text, JSON and so on.

5. After creating the topic finally start the default Kafka producer and link it to this topic that is `test11` in our case.

   ```
   kafka-console-producer.bat --broker-list localhost:9092
   --topic test11
   ```

6. Start typing on the console of the producer program shell to send data to the topic.

![Anaconda - kafka-console-producer.bat --broker-list localhost:9092 --topic test11 console screenshot]

As you can see previously, the same sentences are printed on the console. For our sample, analytic application this would represent a set of tweets.

> In a real world applications, we won't be using this producer but there would be another program that would actually use the Twitter API and fetch the tweets from Twitter and push it on the topics of Kafka.

- **Maven Java project for Spark Streaming**: Integrating Kafka and Spark Streaming can sometimes give problems specially with jar mismatch issues. Due to this we have a sample skeleton maven that can download the jars needed for Spark Streaming and Java integration.

  1. To setup this project, first import the project into your IDE (in our case we used eclipse).

2. Next run the sample `SparkStreamingError.java` from your IDE. Let's go through the code of this sample. This will also serve as a first sample on Spark Streaming and Kafka for us. In this sample example we will read a set of sentences or log messages from Kafka and we would filter out the messages that have the word 'exception' in them.

   First some boiler plate code to build Spark configuration:

   ```
   SparkConf conf = ...
   ```

3. Next, we build the `JavaSparkContext` using this configuration:

   ```
   JavaSparkContext sc = new JavaSparkContext(conf);
   ```

4. Since we are going to use Spark Streaming we need the context specific to Spark Streaming as the main entry point for our application. So we build the Spark Streaming context and here we specify the batch interval based on which the data will be pulled from the source of data on a regular basis. In our case the batch interval is two seconds (its given in micro seconds in the `StreamingContext` constructor):

   ```
   JavaStreamingContext ssc = new JavaStreamingContext(sc, new Duration(2000));
   ```

5. For our sample program we are going to the use the Kafka topic that we created earlier that is `test11` and we are going to read the data from it. The topics of Kafka from which data has to be read are stored in a variable of type `Set`. We also create parameters for Kafka connection and store them in a key value pair object or `Map` object.

   ```
   Set<String> topics = Collections.singleton("test11");
   Map<String, String> kafkaParams = new HashMap<>();
   kafkaParams.put("metadata.broker.list", "localhost:9092");
   ```

   As you can see previously the main parameters to connect to Kafka are the location of the Kafka broker that is the hostname and port on which it is running.

The Spark Kafka integration jar contains a class `KafkaUtils` using which you can directly integrate Kafka with Spark Streaming. It helps you build a connection with the Kafka broker. Here we just need to provide the necessary parameters like streaming context, the type of data transferred, `kafkaparams` containing the broker host and its post and the topic name. The outputs received from this stream of data is a `DStream` object and we call it `directKafkaStream` and this is essentialy a sequence of RDDs. This stream of data is filled up RDD objects and each RDD is essentially a paired RDD filled with a tuple.

```
JavaPairInputDStream<String, String> directKafkaStream =
KafkaUtils.createDirectStream(ssc,
        String.class, String.class, StringDecoder.class,
StringDecoder.class, kafkaParams, topics);
```

6. As we mentioned at the start of this section, we will filter out text that contains the word `exception`. We will go over the content of this `DStream` object and extract each RDD from it. For this we will invoke a `foreach` method on this `DStream` object and within the Java `lambda` function we will first print the number of records in each RDD that we iterate through.

```
directKafkaStream.foreachRDD(rdd -> {
System.out.println("Number of records in this rdd : " +
rdd.count() + " records");
```

7. After printing the number of records in each RDD we will filter the content of this individual RDD. For this we will extract the sentences (log message) from this RDD. As we said the content of RDD is a tuple so we extract the actual content of the log message and store it in a variable `rowLine`. Next, we check that this log message contains the word `exception` in it. If it does, then we return `true` else `false`.

```
rdd.filter(record -> {
String firstLine = record._1;
String rowLine = record._2;

System.out.println("rowLine ------> " + rowLine);
return rowLine.contains("exception");
```

8. Finally we print each log message that was filtered out. For this we just invoke `foreach` on the filtered RDD shown as follows and print the content in it.

   ```
   }).foreach(s ->
   System.out.println("\n Error Text =>" + s._2 + "\n"));
   });
   ```

9. Before the Spark Streaming program starts collecting the data you will have to explicitly invoke the `start` method on the `StreamingContext` object. We also need to wait for this computation to terminate and hence we invoke `awaitTermination` on the `StreamingContext` object.

   ```
   ssc.start();
   ssc.awaitTermination();
   ```

   Now, we are all set let's run this program and type some text on the producer to mimic log messages shown as follows:

   ```
   Anaconda - kafka-console-producer.bat --broker-list localhost:9092 --topic test11
   C:\kafka_2.11-0.9.0.0\bin\windows>kafka-console-producer.bat --broker-list localhost:9092 --topic test11
   Cache is cleared
   Hit count on the server is 50
   Nullpointer exception in class BBCTestService at line 140
   ```

10. As you can see previously we typed three messages on the console and only one has the word `exception` in it. Finally, this will print the output on our Spark Streaming program console.

```
2017-06-08 08:44:06 INFO    JobScheduler:54 - Added jobs for time 1496925846000 ms
2017-06-08 08:44:06 INFO    JobScheduler:54 - Starting job streaming job 1496925846
--- New RDD with 1 partitions and 1 records
2017-06-08 08:44:06 INFO    SparkContext:54 - Starting job: foreach at SparkStreami
2017-06-08 08:44:06 INFO    DAGScheduler:54 - Got job 23 (foreach at SparkStreaming
2017-06-08 08:44:06 INFO    DAGScheduler:54 - Final stage: ResultStage 23 (foreach
2017-06-08 08:44:06 INFO    DAGScheduler:54 - Parents of final stage: List()
2017-06-08 08:44:06 INFO    DAGScheduler:54 - Missing parents: List()
2017-06-08 08:44:06 INFO    DAGScheduler:54 - Submitting ResultStage 23 (MapPartiti
2017-06-08 08:44:06 INFO    MemoryStore:54 - Block broadcast_23 stored as values in
2017-06-08 08:44:06 INFO    MemoryStore:54 - Block broadcast_23_piece0 stored as by
2017-06-08 08:44:06 INFO    BlockManagerInfo:54 - Added broadcast_23_piece0 in memo
2017-06-08 08:44:06 INFO    SparkContext:54 - Created broadcast 23 from broadcast a
2017-06-08 08:44:06 INFO    DAGScheduler:54 - Submitting 1 missing tasks from Resul
2017-06-08 08:44:06 INFO    TaskSchedulerImpl:54 - Adding task set 23.0 with 1 task
2017-06-08 08:44:06 INFO    TaskSetManager:54 - Starting task 0.0 in stage 23.0 (TI
2017-06-08 08:44:06 INFO    Executor:54 - Running task 0.0 in stage 23.0 (TID 23)
2017-06-08 08:44:06 INFO    KafkaRDD:145 - Computing topic test11, partition 0 offs
2017-06-08 08:44:06 INFO    VerifiableProperties:68 - Verifying properties
2017-06-08 08:44:06 INFO    VerifiableProperties:68 - Property group.id is overridd
2017-06-08 08:44:06 INFO    VerifiableProperties:68 - Property zookeeper.connect is
RowLine ------> Nullpointer exception in class BBCTestService at line 140
2017-06-08 08:44:06 INFO    Executor:54 - Finished task 0.0 in stage 23.0 (TID 23)
```

As you can see previously the ellipses contain the text that was printed on the console.

With this we come to an end to our section on our setup for Spark Streaming and Kafka integration. Let's now get into the real world by trying our some real world use cases and seeing the power of Spark Streaming in action.

Trending videos

One of the popular features of Netflix is the trending videos. Using this feature, you can see the top trending movies or videos in the last few hours. Similar features are present on other websites as well for example to show what songs people are currently listening or currently trending hash tags. Usually these type of features require aggregating the stats for the feature over a certain time window. In the case of trending videos you might also have to then find the max hit videos out of those aggregated lists. In the real case scenario Netflix will be doing a much more complex operation but we will keep things simple for the purpose of our small case study here.

Recall that we mentioned a feature from Spark Streaming called as **windowed calculations** which allows us to collect data in windows of predefined time intervals and then we can further work on the data contained in those windows.

A windowed operation can be explained as shown in the following image:

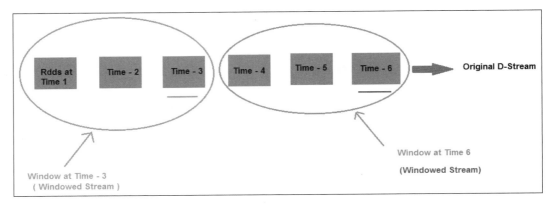

As we have seen in the preceding image a window represents data in a certain period of time. So, if we say data in the last one hour window that means we want data between the current time and the last one hour. Within this timespan there might have been many mini batches of RDDs generated, but we are only interested in the sequence of RDDs that belong to this time window. This is also depicted in the preceding image where we are currently sequencing three sets of RDDs in a window that ends at **Time-6** and similar another window that ends at **Time-3**. Similarly, as time progresses forward there could be multiple windows generated at say **Time-7**, **Time-8** and so on in that order.

For the purpose of this simple case study we assume that we have a source that gives us a video ID and its hit count. So we will aggregate the hit counts of videos within a window of an hour and find the videos with the most hits. This will become the most trending videos for the users as Trending Now. Of course you can do more fancy things on this by applying user recommendations on top of these videos to figure out the video's that the users might like out of these videos and show those videos list first to the users.

We will print this data as comma separate text on the **kafka producer console** and at the same time read this data from the Spark Streaming program. The data on the **kafka producer console** will look as shown below:

Video-1, 1	Video ID and hit count (comma separated)
Video-2, 2	Video ID and hit count (comma separated)
Video-3, 5	Video ID and hit count (comma separated)

The screenshot for the same data when printed on the command console on windows is shown as following:

```
C:\kafka_2.11-0.9.0.0\bin\windows>kafka-console-producer.bat --broker-list localhost:9092 --topic test11
video-1,3
video-2,1
video-3,5
video-4,1
video-5,2
video-6,7
video-3,2
video-6,1
video-4,2
video-4,1
video-3,5
video-2,1
video-1,3
video-5,2
video-3,2
video-6,7
```

Now its time to consume these messages from our Kafka topic in real time and run some analytics on that data. We will create the Spark program for streaming data using the Spark Streaming Context:

```
SparkConf conf = ...
SparkSession spark = ...
JavaSparkContext sc = new JavaSparkContext(spark.sparkContext());
```

After building the `JavaSparkContext` we will now build the `StreamingContext`. It is here that we provide the sliding interval. This is the time that will be used to build the RDDs, so after every interval of this time RDDs will be build and attached to a DStream (sequence). For our local testing purposes we keep the sliding interval as 10 seconds or 10000 milli seconds.

```
JavaStreamingContext ssc = new JavaStreamingContext(sc, new Duration(10000));
```

To keep the output less verbose we mark the `LogLevel` as `ERROR` so that only error logs are shown on the console.

```
sc.setLogLevel("ERROR");
```

Next, we provide the Kafka parameters with the topic names and the broker details:

```
Set<String> topics = Collections.singleton("test11");
Map<String, String> kafkaParams = new HashMap<>();
kafkaParams.put("metadata.broker.list", "localhost:9092");
```

We will now get the DStream that holds all our RDDs collected and to get hold of this `DStream` object we use the utility class `KafkaUtils` provided by the Kafka itself. Using the `KafkaUtils` class, we can invoke a `createDirectStream` method and create a `DStream` by providing it with the Kafka parameters (broker host, port and so on).

```
JavaPairInputDStream<String, String> directKafkaStream =
KafkaUtils.createDirectStream(ssc,
      String.class, String.class,
StringDecoder.class, StringDecoder.class,
kafkaParams, topics);
```

As we showed in the preceding image DStream is built with sequence of RDDs collected at regular intervals (sliding time). For windowed operations we will now create a windowed stream on top of this original DStream. To do that we will invoke the window method on top of the original `DStream` object and provide the duration of the window. In our case we keep the duration less as 30 seconds as it is easy to test it on local machine. So this window time will tell us the video IDs and their hit counts in the last thirty seconds.

Creating the window stream is easy and we can just invoke the window methods on the original `DStream` object and provide the window interval and in our case it is thirty seconds.

```
JavaPairDStream<String, String> windowStream = directKafkaStream.
window(new Duration(30000));
```

Next our task is to analyze the RDDs that come within a window span. To do this we will go over each `rdd` contained in a window and for this we will invoke the `foreach` method on the windowed `DStream` object.

```
windowStream.foreachRDD(rdd -> {
```

To understand which window we are analyzing we will print the ID of the window by using the method `id()` on the `rdd`

```
System.out.println("--- WINDOW ID --- " + rdd.id());
```

We will now extract the strings (rows of data) contained within the `rdd` and for this we will invoke the `map` method on the `rdd`. Since the `rdd` is a paired `rdd`, it has key and value pairs. Within the map `lambda` function we will extract the value from the paired `rdd` and we will split it to extract the video ID and the corresponding video count. We will also store these values in a java value object called as `VideoVO` and return this object from the map `lambda` function. So now our original `rdd` within the DStream is converted to an new `rdd` containing these POJO's called as `VideoVO`.

```
JavaRDD<VideoVO> videoRDD = rdd.map(s -> {
String[] rowStr = s._2.split(",");
VideoVO tvo = new VideoVO();
   tvo.setVideoID(rowStr[0]);
   tvo.setVideoCount(Integer.parseInt(rowStr[1]));
   return tvo;
});
```

To use the power of Spark SQL and simplify our code, we will convert our `rdd` of POJO's to a dataset object. We will invoke the `createDataFrame` method on the spark session object for that.

```
Dataset<Row> videoDS = spark.createDataFrame(videoRDD.rdd(), VideoVO.class);
```

Once our dataset is built, we register it as a temporary view called `videos` and now this is ready to fire some queries on top of it.

```
videoDS.createOrReplaceTempView("videos");
```

Finally, we fire our `spark.sql` query on top of our temporary view and this time we do a group by query. We do a group by on the `videoID` and a sum on the count of the video hits. This is all done as part of the data we obtained within that window. We also sort the results in descending order so videos with maximum hits line up first.

```
spark.sql("select videoID,sum(videoCount) videoHitsCount from videos
group by videoID order by videoHitsCount desc").show();
```

As you can see in the preceding query we also invoked a `show` method on the result of the query and this would print our videos with high hits counts and in that order as shown next:

```
--- WINDOW ID --- 49
+-------+--------------+
|videoID|videoHitsCount|
+-------+--------------+
+-------+--------------+

--- WINDOW ID --- 61
+-------+--------------+
|videoID|videoHitsCount|
+-------+--------------+
|video-3|             5|
|video-1|             3|
|video-4|             1|
|video-2|             1|
+-------+--------------+

--- WINDOW ID --- 73
+-------+--------------+
|videoID|videoHitsCount|
+-------+--------------+
|video-6|             7|
|video-3|             7|
|video-5|             4|
|video-1|             3|
|video-2|             1|
|video-4|             1|
+-------+--------------+
```

As you can see previously our the results printed on the console show three different windows and in each window we can see the videos with high hit counts lined up first. This resulting data can be stored in the database or sent to the UI to be shown to the customers.

Next in our final case study for real-time analytics we will see how we can apply predictive analytics in real time.

Sentiment analysis in real time

We previously covered sentiment analysis and showed how we could train a model over an existing set of data of tweets and later reuse the same model on a different set of data to predict the sentiment whether positive or negative from the words within the tweet. We will now try to do the same sentimental analysis except that at this time we will do it in real time, so we will write a program that will wait for arrival of tweets data on a Kafka topic. When the data is available it will be read by a Spark Streaming program and processed.

Spark provides an out of box feature by which a pretrained model can be saved in external storage.

This trained model can be pulled from external storage (which can be on HDFS) and rebuilt. After the model object is available it can be reapplied on the new set of data and predictions can be made on it. In the new Spark API even the entire pipeline or workflow can be persisted to external storage after it is trained on as set of data. We will be using this feature of Apache Spark to store our pretrained sentimental analysis pipeline to external storage.

We will not be covering sentimental analysis code here, as we already covered that as part of *Chapter 6, Naive Bayes and Sentiment Analysis* when we discussed the Naive Bayes machine learning algorithm.

We have the sentimental analysis pipeline that is pretrained with a data of tweets. The dataset contained tweets that were prelabeled for positive or negative sentiments. To recap our training data is like shown in the following table:

Tweet text	Sentiment (1 for positive and 0 for negative)
The Da Vinci Code book is just awesome.	1
What was so great about the Da Vinci Code, tell me ?	0
I loved the Da Vinci Code !	1

As shown in the following code we first build the pipeline with the different steps in the pipleline workflow entered as stages of the pipeline. To refer to these stages please check the *Chapter 6, Naive Bayes and Sentiment Analysis* on the Naive Bayes model where this is mentioned in detail.

```
Pipeline p = new Pipeline();
p.setStages(new PipelineStage[]{ tokenizer, stopWrdRem, hashingTF, idf,nb});
```

Once our pipeline is read we fit it on the training data so as to train our pipeline and the output of the training that is the trained model is the `PipelineModel`:

```
PipelineModel pm = p.fit(training);
```

Once this model is trained or pretrained it can be stored in an external storage by simply invoking the `save` method on the model and providing the HDFS location

```
pm.save(<LOCATION_IN_HDFS>);
```

 We would recommend trying this example on an environment with good Spark or Hadoop setup like Cloudera environment, or MapR Sandbox. If you run it on Windows you can run into some issues especially while saving the model to external storage; it will require Hadoop libraries on the classpath to do this.

Now, our model is pretrained and saved to external storage and it can be rebuilt again from that location. Let's now work on actual Spark Streaming program. You will be surprised as to how short yet how powerful this program is.

As usual, we will first build our boiler plate code for initiating the Spark configuration and `SparkSession`. To maintain brevity we are not showing the full code for these boiler plate code below.

```
SparkConf conf = ...
SparkSession spark = ...
```

We will build the `JavaSparkContext` using the `spark` session object.

```
JavaSparkContext sc = new JavaSparkContext(spark.sparkContext());
```

As this is a Spark Streaming program that gathers data at real-time, we will next build the `JavaStreamingContext` and provide the time interval after which next micro batch of data will be pulled from the stream. We are going to pull the data from a Kafka topic:

```
JavaStreamingContext ssc = new JavaStreamingContext(sc, new Duration(10000));
```

Next, we load our model from external storage and store it in our `PipeLineModel` object. Thus, our `pm` instance now has our pretrained model that we can use and apply on our dataset.

```
PipeLineModel pm = pm.load(<HDFS_LOCATION>);
```

We have to provide the necessary parameters using which we can stream the data from the Kafka topic as a DStream. To do this we first create a set of topics:

```
Set<String> topics = Collections.singleton("test11");
```

Next we provide the `KafkaParams` containing the broker host and port.

```
Map<String, String> kafkaParams = new HashMap<>();
kafkaParams.put("metadata.broker.list", "localhost:9092");
```

Using the `KafkaUtils` class, we finally pull the `DStream` out of the Kafka topics at intervals we set in the streaming context.

```
JavaPairInputDStream<String, String> directKafkaStream =
KafkaUtils.createDirectStream(ssc,String.class, String.class,
StringDecoder.class, StringDecoder.class, kafkaParams, topics);
```

Once you have received the first mini batch of data in DStream you can now go over the items of the DStream using a `foreachRDD` method. Each item in the DStream is an paired `rdd` (that is it is a tuple object) and from it we can extract the actual tweet. Within the `lambda` function, we first print the details about the `rdd` and the amount of data in it in terms on number of tweets.

```
directKafkaStream.foreachRDD(rdd -> {
System.out.println("--- New RDD with " + rdd.partitions().size()
   + " partitions and " + rdd.count() + " records");
```

Next, we invoke a `map` function on each of this paired `rdd` so that the map gets invoked on every element that is stored within the `rdd`. Within this `map` function we first pull out the tweet text and this is present in the second element of the tuble as shown by s._2 in the code below. Next we populate this value in a `TweetVO` object. Finally we return the `TweetVO` object from this map `lambda` function. So the output `rdd` is now a distributed collection of `TweetVO` objects.

```
JavaRDD<TweetVO> tweetRdd = rdd.map(s -> {
String rowStr = s._2;
  TweetVO tvo = new TweetVO();
    tvo.setTweet(rowStr);
    return tvo;
});
```

We will now convert this `rdd` of `TweetVO` objects to a dataset object.

```
Dataset<Row> tweetsDs = spark.createDataFrame(tweetRdd.rdd(), TweetVO.
class);
tweetsDs.show();
```

We also invoked the `show` method on our preceding dataset and this would print out the first few lines available in our micro batch of tweet messages:

Finally, we now go ahead and apply our pre-trained model on the tweets that we received using streaming. For this, we will invoke the `transform` method on the `tweets` dataset that we collected using streaming.

```
Dataset<Row> predictedResult = pm.transform(tweetsDs);
```

The result of this transformation is also stored in the form of a dataset. Let's now see how well our `predicted` model did. We can see the first few lines in our predicted dataset by invoking a `show` method on it as shown next.

```
predictedResult.show();
```

This would print the result as:

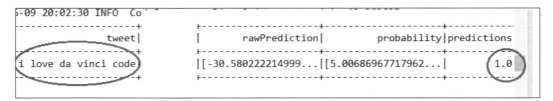

We have omitted some columns in the preceding image but the main columns are the first column which shows the actual message and the last column which shows the predicted sentiment. In this case the sentiment is positive or **1**.

Chapter 12

Apart from these we can also sent a negative sentiment containing text from the **Kafka producer console**. For example, we can send a text as da vinci code is bad and as we can see in the image below the sentiment would be predicted as negative or **0**.

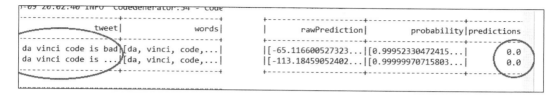

You got your result in the dataset format. The users can also push this dataset to external storage. For simplicity, we are pushing the results in parquet format to HDFS. Even though this approach looks simple yet it is a very good approach simply because parquet is very fast due to its columnar nature and compression also works very well on parquet.

```
predictedResult.write().format("parquet").save(<HDFS LOCATION>);
```

> Even though we have stored the results as parquet to HDFS the results can also be stored in a NoSQL database like Cassandra or HBase and further evaluated.

As we mentioned earlier for Spark Streaming the actual computation only start when you specifically invoke the start method on the streaming context so next we finally invoke the start method on the Spark Streaming context.

```
ssc.start();
```

We also invoke awaitTermination method on the Streaming context for proper termination of the program.

```
ssc.awaitTermination();
```

Thus sentimental analysis can be easily applied on streaming data using Spark Streaming. With this we come to the end of this chapter.

Summary

In this chapter, we learnt about real-time analytics and saw how big data can be used in real-time analytics apart from batch processing too. We introduced the product Impala that can be used to fire fast SQL queries on big data which is usually stored in Parquet format in HDFS. While looking at Impala we briefly did a simple case study on flight analytics using Impala. We later covered Apache Kafka a messaging product that can be used in conjunction with big data technologies and build real time data stacks. Kafka is a scalable messaging solution and we showed how it can be integrated with Spark Streaming module of Apache Spark. Spark Streaming let's you collect data in mini batches in real time and it calls sequence of these mini batches as streams. Spark Streaming is becoming very popular these days as it is a good scalable solution that fits into the needs of many users. We finally covered a few cases studies using Apache Kafka and Spark Streaming and showed how complex use cases like real time predictive analysis can be easily done using the API's of these products.

In the next chapter we will cover a module which becoming very hot these days and it is called as deep learning.

13
Deep Learning Using Big Data

In recent years, if there is something that has gained lot of traction and advancement in the field of computer science research it is **deep learning**. If you pick up any of the latest research papers, you will see that a lot of them are in fact in the field of deep learning only. Deep learning is a form of machine learning that sat idle for quite some time, until recently, when computations on multiple parallel computers became more advanced. The technology behind the self-driving car or an ATM recognizing a hand-written check is all done through deep learning in real life. So, what exactly is deep learning? We will cover the basic details of deep learning in this chapter. It is a form of machine learning that roughly mimics the working of a human brain using neural networks. Deep learning is a vast field and a growing one too, so this chapter should serve as a bare minimum for anyone trying to find a basic introduction on the topic. More advanced descriptions are beyond the scope of this chapter. However, unsurprisingly, many resources are now available on this topic for free on the internet.

In this chapter, we will cover the following topics:

- Introduction to neural networks
- Perceptron and sigmoid neurons
- Multi-layer perceptrons
- How to improve the performance of neural networks
- How deep learning and neural networks mingle
- Use cases and advantages of deep learning
- First sample case study of flower classification
- Deeplearning4j library
- Where to find extra information in deep learning

Introduction to neural networks

Our human brain has millions of neurons that talk and transfer signals to each other. So, when one neuron calculates a signal and transfers it to another neuron, the second neuron that is connected to the first neuron becomes their input and acts on it. In this way, the initial input goes through various neurons while being altered at each level until a final deduction can be made. You can think of our brain as a graph of these neurons interconnected to each other, and sending signal or inputs to each other.

Let's see how a typical neuron in the human brain looks:

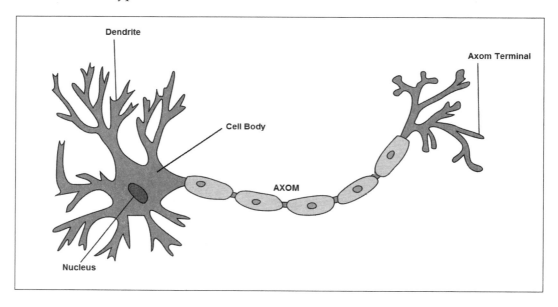

This picture of a human neuron has been taken from Wikipedia and can be seen at this link: https://en.wikipedia.org/wiki/Dendrite. The neuron has some important components, as seen in the labels in the image. They are explained as follows:

- **Dendrite**: These are hair-like parts. They connect the neuron to other neurons, and they are used in taking input from other neurons.
- **Cell Body**: This is the place where the input received by dendrites is acted upon.
- **Axom Terminal**: The output that is generated from this neuron is transmitted out from the axom terminals to other neurons.

In a human brain, we have millions of neurons like these. The neurons take in input from many other neurons and process this input to generate an output that is transmitted further. This process continues till an outcome is reached.

In an **artificial neural network** built using computers, the same approach of our human nervous system is replicated. So, we have a computer program that mimics a human neuron. We call it an artificial neuron. It is depicted in the following diagram:

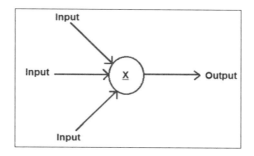

As seen in the previous image, the circle represents an artificial neuron that takes in input from various sources and performs processing on it (represented by **X**), before finally generating an output. This artificial neuron is nothing but a computer program or algorithm that is doing this functionality based on certain criteria, as we will see further on in this chapter.

A mesh of such neurons connected together can mimic the working of our nervous system, shown as follows:

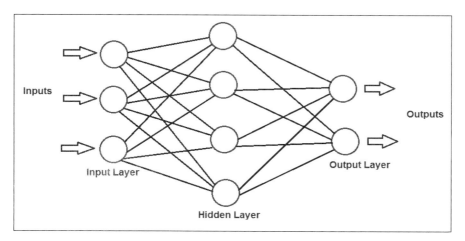

As shown in the previous image, there are many artificial neurons connected to each other, transferring their computed result based on their input to other neurons. As the input progresses through many neurons, an outcome is reached. We will be covering the neural network later on in this chapter.

These artificial neural networks can be trained on historical data and, based on that data; they can build up their knowledge base or set of rules and store them internally. Based on this knowledge, they can make a prediction when given a new piece of data. For example, if we have historical data for the stock prices of various stocks from stock exchanges over the past few years, we can train a neural network with this data. Upon training, this neural network can then make guesses about the stock price of a stock based on various features.

As we mentioned previously, there is one vast difference between the artificial neural networks and the human brain, that is, the human brain can work without data and create new stuff, which is underlined as the creativity of humans. This is something which is not possible (at least not for now) with artificial neural networks, as they only work with historical data.

There are many ways in which an artificial neuron can be represented. We will be concentrating mainly on a perceptron, which is one way of representing artificial neurons.

Perceptron

A perceptron is a type of artificial neuron that is mathematical and programmatic. It takes in many inputs and applies weights to them based on the importance of the inputs, and then adds a bias before using this mathematical approach to figure out a result. This result from the perceptron is then fed to a machine learning algorithm, such as logistic regression. We call this algorithm as an activation function, which is then is used to predict the final result of the outcome.

The perceptron is depicted as follows:

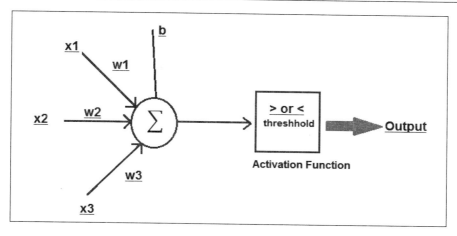

As you can see in the previous image, a perceptron depicts an artificial neuron that takes in various inputs in binary form and multiplies them with a weight, **w**. The weight is calculated based on the importance of the input. A bias value is also added, along with the weights. Now, the entire combination is summed up by the perceptron. Finally, the summed-up output is tested against a threshold value, and we call this as an **Activation Function**. If the value is above a threshold, a deduction is made. The bias is added to normalize the sum of the results of the weight and input multiplication, which helps the perceptron to either go above or below the threshold.

As the output of the perceptron is a binary output, a perceptron can be used as a good linear binary classifier. Let's look at a very basic example of how a perceptron works.

Suppose you want to look at admission into the computer science department of a university. Your criteria are based on the following questions:

- How knowledgeable are the professors?
- Are extensive courses available on artificial intelligence?
- Is the campus good?

Now suppose you have three data items for these based on three different universities:

University	Good Professors	Courses on AI	Good Campus
College - A	1	1	0
College - B	0	1	1
College - C	0	0	1

Deep Learning Using Big Data

Let's assign the weights to the different features

- Good Professors = 4
- Courses on AI = 3
- Good Campus = 2

As you can see, the presence of good professors takes the highest priority in our case.

Now suppose our minimum threshold is five, and our bias is one, then the calculated value of perceptrons for all the inputs based on our formula is:

$$\text{Calculation Output} = b + \sum_{k=0}^{n} w * x$$

Where b is the bias and the threshold needed for perceptron to fire (that is, the perceptron takes its decision based on this value), w is the weight for each input, and x is the actual input value.

And:

$$\text{Perception result} = 0 \text{, if Calculation Output} \leq 0$$

$$result = 1 \text{, if Calculation}$$

So, as seen in the equation for the result previously, the perceptron result is positive, or 1. If the calculation result is greater than zero, it is zero. After going over the mathematical formula, let's now apply this formula on our sample case study to check which college the student would prefer to join. Again, our bias or threshold value is -5.

University	Good Professors	Courses on AI	Good Campus	Calculation	Result	Perceptron Output
College - A	1	11	0	-5 + (4 * 1 + 3 * 1 + 2 * 0)	3	1
College - B	0	1	1	-5 + (4 * 0 + 3 * 1 + 2 * 1)	0	0
College - C	0	0	1	-5 + (4 * 0 + 3 * 0 + 2 * 1)	-3	0

As we said previously, the threshold is 5, so in our case only in the first case, that is, **College - A** does this meets the criteria (as the decision in the perceptron output is positive). The student would therefore choose the computer course from this college. As you can see, there is too much dependency on the first criteria of good professors. If we lower the threshold to 4 then **College - B** would also be selected. Thus, as you can see, even if we change the threshold by a small amount, it can have a big impact on the outcome.

In short, a perceptron is a binary linear classifier that will classify the input into one of the two binary categories. It will segment the result based on whether the value is greater or lesser than the threshold; in our test case, the student would either select the college or reject it. At this point, it looks like this base perceptron is a great decision classifier, and in fact for many years after its inception, a lot of people thought perceptrons to be a magic bullet that solved our artificial intelligence problems. However, perceptrons have their own set of problems, which we will discuss shortly.

Problems with perceptrons

A small change in the threshold value can have a big impact on the output of the perceptron. As seen previously, if we change the threshold to just 4 from 5, that is, we changed it by just 1 unit, we were able to select an extra college for the student. This is a problem with perceptrons, and to handle this we use a sigmoid neuron.

The perceptron can be used to represent logical functions. We will show two simple examples of how a perceptron can be used to represent two simple logical functions.

- **Logical AND:** If a perceptron has two input weights as 1 and 1 and we enter a bias value as 1.5, then only when both the inputs that are passed to the perceptron are 1 and 1 is the output positive. The following figure depicts an AND function using a perceptron:

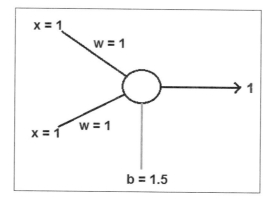

We can test the perceptron with the following inputs. As you can see, it only fires when both the inputs are 1:

Input - 1	Input - 2	Calculated Value	Output
1	0	-0.5	0
0	1	-0.5	0
1	1	0.5	1

- **Logical OR**: Similar to logical AND, a logical OR can also be implemented, as follows:

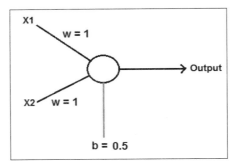

We can pass many inputs to this perception, but the output will only be zero when both the inputs are zero. Let's look at the input and output in the following table:

Input - 1	Input - 2	Calculated Value	Output
1	0	0.5	1
0	1	0.5	1
1	1	1.5	1
0	0	-0.5	0

Here, the perceptron would fire even if only one of the inputs is 1.

As you have seen, a perceptron can be used to represent simple logical functions. This made it quite popular when it was first invented. However, a famous research paper came out of MIT where the writers showed how a perceptron cannot be used to represent a XOR function. To build a XOR function you would have to use multiple perceptrons instead of just a single one, and it is impossible to represent a XOR with a single perceptron. We are not showing the XOR function here, but we would urge readers to check out the details of how an XOR function is represented using perceptrons on the web.

We have discussed two solutions to the afore mentioned drawbacks; one being the usage of sigmoid neurons instead of plain perceptrons, and the other being using a combination of perceptrons instead of a single one. Next, we'll see what a sigmoid neuron is.

Sigmoid neuron

We previously discussed the drawbacks of perceptrons and how they change their output tremendously with subtle changes, which happens when it uses a linear activation function (that is, it just checks output based on a threshold value). So, the relation between input and output is steep. This is a problem that hinders the learning capability of a perceptron, particularly if it is to be used in a machine learning model. To make a perceptron successful as a learner, we have to make subtle changes in its input and record its output only on the basis of that we can figure out at what stage it has learnt the rule well enough to make predictions. If even subtle changes cause vast variations in output, we will not be able to figure out at which stage of variation the perceptron is giving its best prediction.

To fix the issue, instead of using a linear activation function that is based on whether the output is greater or less than the threshold, a different activation function can be used. One of the most popular functions available is the sigmoid function. Recall that we studied sigmoid functions in the chapter on logistic regression. This is the same sigmoid function, and it will convert any number fed to it, to fall as a real number between 0 and 1; for example, 0.675 or 0.543. Sigmoid is used as an activation function on the output of the perceptron and it will convert the output to a real number between 0 and 1. The formula for the sigmoid function is

$$f(x) = \frac{1}{1+e^{-x}}$$

In this formula:

- e: the natural logarithm base
- x: the calculated value of the perceptron, that is,

$$b + \sum_{k=0}^{n} w * x$$

Thus, as you can see, the sigmoid neuron is nothing but a modified perceptron, and it is this neuron that is extensively used in artificial neural networks. This makes the artificial neuron suited for learning purposes as you can make subtle changes in its input without making extreme changes in its output, thereby avoiding the risk of unnecessarily firing the neuron. There is tremendous use of this in algorithms like gradient descent, where we try to reduce the error in the output by making subtle and continuous changes in the input.

Let's see now how the sigmoid function is used along with a perceptron:

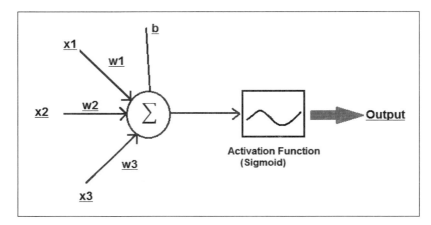

As you can see, the activation function now used is a sigmoid function. This causes the output of the perceptron to fall into the range of 0 and 1 as real numbers.

As sigmoid neurons are better in performance for learning algorithms, they can be combined with multiple sigmoid neurons to form complex learnings systems. These can be used in various use cases for machine learning activities, such as handwritten digit recognition, voice recognition, and so on.

Next, we will look into the concept of multiple artificial neurons.

Multi-layer perceptrons

Several sigmoid neurons can be connected and stacked in layers. These layers can together take input and generate output. The output from these layers can be fed further to other layers, which can be continued till you reach a layer that finally generates output. In each layer, a new deduction can be made or each layer helps in enhancing the learning capability of an artificial neural network. This set of sigmoid neurons connected in layers is called a multi-layer perceptron. The name is misleading, as these are essentially sigmoid neurons connected together.

There are various ways in which these artificial neurons can be connected. For example, the following image shows a multi-layer perceptron network, and this is also called a feed forward network.

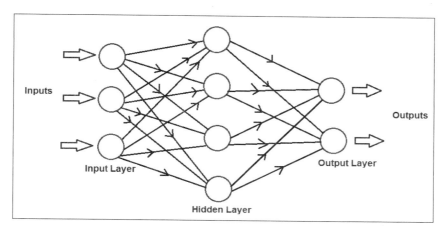

As seen in the previous image, this is a simple example of a single hidden layer multi-perceptron network. This is also called feed forward because the flow of computations is only going in one direction, that is, forwards. There are other kinds of networks where the output is fed back to the neurons and so the direction of flow is in two directions; such networks are called recurrent neural networks and they are beyond scope for this introductory chapter.

Even though the preceding network roughly depicts a human brain or human neural network system (as the human system is much more complex), it is still very useful and is used in many practical applications. This type of network can be used in different forms of machine learning techniques, such as classifications, regression, or clustering. As seen, the network has one input where the input is fed to the neurons, which is input that is next transferred to the hidden layer. All the neurons in the hidden layer receive input from all the neurons in the input layer. After making their computations on top of the data, the output from the sigmoid neurons of the hidden layer is finally transferred to the output layer, where the finally deduction is made. As the final output is received, recall that the output of the sigmoid neuron is actually the probability with a range between 0 and 1. Thus, if the above network is trying to predict a binary classification, for example, whether a stock should be held or sold, then the two output neurons in the output layer can depict the probability of hold and sold. Finally, classification would depend on which probability is higher.

Predicting a result is one thing, but predicting a good output is a whole different story. The task of a neural network is to predict results that are as close to the actual output as possible. Next, let's see how we can tone the accuracy of our neural networks.

Accuracy of multi-layer perceptrons

Improving the performance of this network so that its output gets closer and closer to the expected result is done on a trial and error basis. The output of this network is take and compare its output against the actual results. The difference between these values gives us the error in our result. As we have seen in previous chapters, this can be represented by mean squared error. To improve the performance of our neural network, we have to come up with a combination of weights and biases which when applied to the various neurons produce an output with minimum mean squared error. The formula for mean squared error is simple and is depicted as:

$$Mean\ Squared\ Error = \frac{1}{2n}\sum_{k=0}^{n}(y(x)-a)^2$$

In the formula, the n stands for the number of inputs, x is the input value, and y is the function applied to the input to get the generated output, and finally, a is the actual value of the result. Mean squared error is nothing but the difference between the generated output and the actual output, and shows the amount of error in the predicted result.

> There are many other ways of finding the error between the actual output and the generated output. But for the purpose of this introductory chapter, we only use mean squared error as it has given good results in many practical applications.

There are mathematical ways by which mean squared errors can be reduced. Two of the most popular ways of reducing the error are gradient descent and stochastic gradient descent.

Gradient descent is a mathematical approach of minimizing the cost of a mathematical evaluation function. In our case, our cost is the mean squared error, so by using gradient descent we would try to minimize this error. The following image shows a graph of the error or the cost function versus the feature (coefficient), which shows the error going up and down. In the case of a neural network, this coefficient will be the weight applied to various inputs. The following image is simplistic as it shows the error value versus just one coefficient, but in real-world applications, the number of inputs can run into the thousands. Due to this, specific approaches like Gradient Descent are a must when evaluating the best weight values:

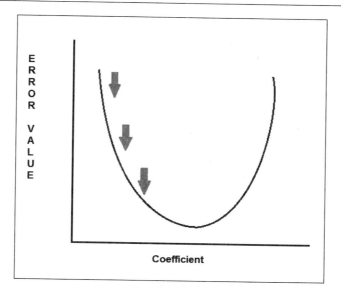

As seen in the previous image, the approach of gradient descent is simple. As we can see, the graph shows how the error value will go down with the change in coefficient, while beyond a certain change the error value will move up again. It is due to this fact that the chart in the image has a step curve. If we want to find the minimum point in this graph, we can increase the value of the coefficient by a minimum amount and see the corresponding drop in error value. We can continue doing this until the error starts rising further, and this way we would hit the minimum point. This is similar to dropping a ball from the top of the chart and seeing how where it would stop at the bottom of the curve. It is also depicted in the arrows in the chart, where we can see how the value will go down. In mathematical terms, a gradient value is calculated (formally called a delta), and this is multiplied by the weights of the input to make a new prediction. The new error is computed. This is done again and again until a minimum value is reached.

There are two ways of working on the dataset with gradient descent:

- **Full batch approach:** In this approach, the entire dataset is used and a minimum value is found. This is called gradient descent and it is a slow process, as every time calculation is done it is on an entire dataset.

- **Partial batch approach:** In this approach, a subset of datapoints is used for calculating the gradient (this is multiplied with the weights to reduce the value of the cost function or error). As this uses only a subset of the datapoints, this is much faster than the normal gradient descent approach. This approach is very popular due to its performance and is called stochastic gradient descent.

The information we have provided in this chapter is very basic. We will tell you at the end of the chapter where to find more information on artificial neural networks.

We have discussed how we can build a multi-layer neural network and optimize it to bring out good predicted results. However, we have not yet covered what exactly deep learning is.

Deep learning

In the last section, we saw how a number of perceptrons can be stacked together in multiple layers to start a learning network. We saw an example of a feed forward network with just one hidden layer. Apart from just a single hidden layer, we can have multiple hidden layers stacked one after the other. This would enhance the accuracy of the artificial neural network further. When an artificial neural network has multiple hidden layers (that is, greater than one), this approach is called deep learning as the network is deep.

Deep learning is currently one of the most widely studied research topics and it is practically used in many real-world applications.

Let's now see some of the advantages and real-world use cases of deep learning.

Advantages and use cases of deep learning

There are two main advantages of deep learning:

1. **No feature engineering required**: In traditional machine learning, feature engineering is of the utmost importance if you want your models to work well. There are teams of data scientists who spend a great amount of time doing feature engineering to train their models well. From the perspective of neural networks, they automatically learn the features from the data and they do not require any feature engineering.

2. **Accuracy:** The accuracy of neural networks is the main reason why they are so popular. They have a high level of accuracy and increase at a tremendous pace along with the latest research.

It is due to these main advantages that deep neural networks are often used in real-world applications, and some of those applications are:

- Hand digit character recognition
- Time series type predictions, such as predicting weather, stock prices, or share prices
- Self-driving cars
- Data compression
- General classification tasks like disease prediction, stock prediction, and so on

So much for the theory. Let's now try to understand the concept of multi-perceptron using a simple case study.

Flower species classification using multi-Layer perceptrons

This is a simple hello world-style program for performing classification using multi-layer perceptrons. For this, we will be using the famous Iris dataset, which can be downloaded from the UCI Machine Learning Repository at https://archive.ics.uci.edu/ml/datasets/Iris. This dataset has four types of datapoints, shown as follows:

Attribute name	Attribute description
Petal Length	Petal length in cm
Petal Width	Petal width in cm
Sepal Length	Sepal length in cm
Sepal Width	Sepal width in cm
Class	The type of iris flower that is Iris Setosa, Iris Versicolour, Iris Virginica

This is a simple dataset with three types of Iris classes, as mentioned in the table.

From the perspective of our neural network of perceptrons, we will be using the multi-perceptron algorithm bundled inside the spark ml library and will demonstrate how you can club it with the Spark-provided pipeline API for the easy manipulation of the machine learning workflow. We will also split our dataset into training and testing bundles so as to separately train our model on the training set and finally test its accuracy on the test set. Let's now jump into the code of this simple example.

Deep Learning Using Big Data

First, create the Spark configuration object. In our case, we also mention that the master is `local` as we are running it on our local machine:

```
SparkConf sc = new SparkConf().setMaster("local[*]");
```

Next, build the `SparkSession` with this configuration and provide the name of the application; in our case, it is `JavaMultilayerPerceptronClassifierExample`:

```
SparkSession spark = SparkSession
        .builder()
        .config(sc)
        .appName("JavaMultilayerPerceptronClassifierExample")
        .getOrCreate();
```

Next, provide the location of the `iris` dataset file:

```
String path = "data/iris.csv";
```

Now load this dataset file into a Spark `dataset` object. As the file is in an csv format, we also specify the format of the file while reading it using the `SparkSession` object:

```
Dataset<Row> dataFrame1 = spark.read().format("csv").load(path);
```

After loading the data from the file into the dataset object, let's now extract this data from the dataset and put it into a Java class, `IrisVO`. This `IrisVO` class is a plain POJO and has the attributes to store the data point types, as shown:

```
public class IrisVO {
    private Double sepalLength;
    private Double petalLength;
    private Double petalWidth;
    private Double sepalWidth;
    private String labelString;
```

On the dataset object `dataFrame1`, we invoke the to `JavaRDD` method to convert it into an RDD object and then invoke the map function on it. The map function is linked to a `lambda` function, as shown. In the `lambda` function, we go over each row of the dataset and pull the data items from it and fill it in the `IrisVO` POJO object before finally returning this object from the `lambda` function. This way, we get a `dataMap` rdd object filled with `IrisVO` objects:

```
JavaRDD<IrisVO> dataMap = dataFrame1.toJavaRDD().map( r -> {
    IrisVO irisVO = new IrisVO();
        irisVO.setLabelString(r.getString(5));
        irisVO.setPetalLength(Double.parseDouble(r.getString(3)));
        irisVO.setSepalLength(Double.parseDouble(r.getString(1)));
        irisVO.setPetalWidth(Double.parseDouble(r.getString(4)));
```

```
            irisVO.setSepalWidth(Double.parseDouble(r.getString(2)));

            return irisVO;
    });
```

As we are using the latest Spark ML library for applying our machine learning algorithms from Spark, we need to convert this RDD back to a dataset. In this case, however, this dataset would have the schema for the individual data points as we had mapped them to the `IrisVO` object attribute types earlier:

```
Dataset<Row> dataFrame = spark.createDataFrame(dataMap.rdd(), IrisVO.class);
```

We will now split the dataset into two portions: one for training our multi-layer perceptron model and one for testing its accuracy later. For this, we are using the prebuilt `randomSplit` method available on the dataset object and will provide the parameters. We keep 70 percent for training and 30 percent for testing. The last entry is the 'seed' value supplied to the `randomSplit` method.

```
Dataset<Row>[] splits = dataFrame.randomSplit(new double[]{0.7, 0.3}, 1234L);
```

Next, we extract the splits into individual datasets for training and testing:

```
Dataset<Row> train = splits[0];
Dataset<Row> test = splits[1];
```

Until now we had seen the code that was pretty much generic across most of the Spark machine learning implementations. Now we will get into the code that is specific to our multi-layer perceptron model. We will create an `int` array that will contain the count for the various attributes needed by our model:

```
int[] layers = new int[] {4, 5, 4, 3};
```

Let's now look at the attribute types of this int array, as shown in the following table:

Attribute value at array index	Description
0	This is the number of neurons or perceptrons at the input layer of the network. This is the count of the number of features that are passed to the model.
1	This is a hidden layer containing five perceptrons (sigmoid neurons only, ignore the terminology).
2	This is another hidden layer containing four sigmoid neurons.
3	This is the number of neurons representing the output label classes. In our case, we have three types of Iris flowers, hence three classes.

After creating the layers for the neural network and specifying the number of neurons in each layer, next build a `StringIndexer` class. Since our models are mathematical and look for mathematical inputs for their computations, we have to convert our string labels for classification (that is, Iris Setosa, Iris Versicolour, and Iris Virginica) into mathematical numbers. To do this, we use the `StringIndexer` class that is provided by Apache Spark. In the instance of this class, we also provide the place from where we can read the data for the label and the column where it will output the numerical representation for that label:

```
StringIndexer labelIndexer = new StringIndexer().
setInputCol("labelString").setOutputCol("label");
```

Now we build the features array. These would be the features that we use when training our model:

```
String[] featuresArr = {"sepalLength","sepalWidth","petalLength","petalWidth"};
```

Next, we build a features vector as this needs to be fed to our model. To put the feature in vector form, we use the `VectorAssembler` class from the Spark ML library. We also provide a features array as input and provide the output column where the vector array will be printed:

```
VectorAssembler va = new VectorAssembler().setInputCols(featuresArr).
setOutputCol("features");
```

Now we build the multi-layer perceptron model that is bundled within the Spark ML library. To this model we supply the array of layers we created earlier. This layer array has the number of neurons (sigmoid neurons) that are needed in each layer of the multi-perceptron network:

```
MultilayerPerceptronClassifier trainer = new
MultilayerPerceptronClassifier()
        .setLayers(layers)
        .setBlockSize(128)
        .setSeed(1234L)
        .setMaxIter(25);
```

The other parameters that are being passed to this multi-layer perceptron model are:

Block Size	Block size for putting input data in matrices for faster computation. The default value is 128.
Seed	Seed for weight initialization if weights are not set.
Maximum iterations	Maximum number of iterations to be performed on the dataset while learning. The default value is 100.

Finally, we hook all the workflow pieces together using the pipeline API. To this pipeline API, we pass the different pieces of the workflow, that is, the `labelindexer` and vector assembler, and finally provide the model:

```
Pipeline pipeline = new Pipeline().setStages(new PipelineStage[]
{labelIndexer, va, trainer});
```

Once our `pipeline` object is ready, we fit the model on the training dataset to train our model on the underlying training data:

```
PipelineModel model = pipeline.fit(train);
```

Once the model is trained, it is not yet ready to be run on the test data to figure out its predictions. For this, we invoke the `transform` method on our model and store the result in a `Dataset` object:

```
Dataset<Row> result = model.transform(test);
```

Let's see the first few lines of this result by invoking a `show` method on it:

```
result.show();
```

This would print the result of the first few lines of the result dataset as shown:

```
+-----------+-----------+----------+-----------+----------+-----+--------------------+----------+
|labelString|petalLength|petalWidth|sepalLength|sepalWidth|label|            features|prediction|
+-----------+-----------+----------+-----------+----------+-----+--------------------+----------+
|     setosa|        1.0|       0.2|        4.6|       3.6|  1.0|[4.6,3.6,1.0,0.2]|       1.0|
|     setosa|        1.1|       0.1|        4.3|       3.0|  1.0|[4.3,3.0,1.1,0.1]|       1.0|
|     setosa|        1.2|       0.2|        5.0|       3.2|  1.0|[5.0,3.2,1.2,0.2]|       1.0|
|     setosa|        1.4|       0.2|        4.4|       2.9|  1.0|[4.4,2.9,1.4,0.2]|       1.0|
|     setosa|        1.5|       0.2|        4.6|       3.1|  1.0|[4.6,3.1,1.5,0.2]|       1.0|
|     setosa|        1.5|       0.2|        4.9|       3.1|  1.0|[4.9,3.1,1.5,0.2]|       1.0|
|     setosa|        1.5|       0.2|        5.3|       3.7|  1.0|[5.3,3.7,1.5,0.2]|       1.0|
|     setosa|        1.5|       0.4|        5.1|       3.7|  1.0|[5.1,3.7,1.5,0.4]|       1.0|
|     setosa|        1.6|       0.2|        4.8|       3.1|  1.0|[4.8,3.1,1.6,0.2]|       1.0|
|     setosa|        1.6|       0.2|        5.1|       3.8|  1.0|[5.1,3.8,1.6,0.2]|       1.0|
|     setosa|        1.7|       0.2|        5.4|       3.4|  1.0|[5.4,3.4,1.7,0.2]|       1.0|
|     setosa|        1.7|       0.3|        5.7|       3.8|  1.0|[5.7,3.8,1.7,0.3]|       1.0|
|     setosa|        1.7|       0.5|        5.1|       3.3|  1.0|[5.1,3.3,1.7,0.5]|       1.0|
| versicolor|        3.3|       1.0|        5.0|       2.3|  2.0|[5.0,2.3,3.3,1.0]|       2.0|
| versicolor|        4.0|       1.2|        5.8|       2.6|  2.0|[5.8,2.6,4.0,1.2]|       2.0|
| versicolor|        4.0|       1.3|        6.1|       2.8|  2.0|[6.1,2.8,4.0,1.3]|       2.0|
| versicolor|        4.3|       1.3|        6.4|       2.9|  2.0|[6.4,2.9,4.3,1.3]|       2.0|
| versicolor|        4.4|       1.4|        6.7|       3.1|  2.0|[6.7,3.1,4.4,1.4]|       2.0|
| versicolor|        4.5|       1.3|        5.7|       2.8|  2.0|[5.7,2.8,4.5,1.3]|       2.0|
| versicolor|        4.5|       1.5|        5.6|       3.0|  2.0|[5.6,3.0,4.5,1.5]|       2.0|
+-----------+-----------+----------+-----------+----------+-----+--------------------+----------+
```

As seen in the previous image, the last column depicts the predictions made by our model. After making the predictions, let's now check the accuracy of our model. For this, we will first select two columns in our model which represent the predicted label, as well as the actual label (recall that the actual label is the output of our `StringIndexer`):

```
Dataset<Row> predictionAndLabels = result.select("prediction",
    "label");
```

Finally, we will use a standard class called `MulticlassClassificationEvaluator`, which is provided by Spark for checking the accuracy of the models. We will create an instance of this class. Next, we will set the metric name of the metric, that is, accuracy, for which we want to get the value from our predicted results:

```
MulticlassClassificationEvaluator evaluator =
new MulticlassClassificationEvaluator()
    .setMetricName("accuracy");
```

Next, using the instance of this evaluator, invoke the `evaluate` method and pass the parameter of the dataset that contains the column for the actual result and predicted result (in our case, it is the `predictionAndLabels` column):

```
System.out.println("Test set accuracy = " +
    evaluator.evaluate(predictionAndLabels));
```

This would print the output as:

```
17/06/27 19:17:24 INFO DAGSchedule
Test set accuracy = 0.95
17/06/27 19:17:24 INFO SparkUI: St
```

If we get this value in a percentage, this means that our model is 95% accurate. This is the beauty of neural networks - they can give us very high accuracy when tweaked properly.

With this, we come to an end for our small hello world-type program on multi-perceptrons. Unfortunately, Spark support on neural networks and deep learning is not extensive; at least not until now. Due to this, we will use another library that is easily integrated with Spark and has extensive support for different types of neural networks, like feed forward neural networks, recurrent neural network, or convolution neural networks.

Note: The types of neural network we have mentioned above are all part of the Deeplearning4j library.

Let's take a brief look at the Deeplearning4j library.

Deeplearning4j

This is a Java library that is used to build different types of neural networks. It can be easily integrated with Apache Spark on the big data stack and can even run on GPUs. It is the only main Java library out there currently that has a lot of built-in algorithms focusing on deep learning. It also has a very good online community and good documentation, which can be checked on its website at `https://deeplearning4j.org`.

There are lots of submodules within this Java library and we need some of those sub modules for running our machine learning algorithms. To check out more detail and running samples within Deeplearning4j, please refer to their documentation. We will not cover Deeplearning4j API in this book, please refer to `https://deeplearning4j.org` for more information on its documentation.

In order to generate the curiosity of the reader as to what all can be accomplished with deep learning we will end the chapter with another simple sample case study of hand written digit recognizition using neural networks. We will explain the concepts involved along with the code.

 As this is an introductory chapter we won't be covering all the concepts in detail in the the next section. We would urge the users to refer to the book "Deep Learning" from MIT press, for more information on deep learning. The professors who wrote this book have been kind enough to give this book for free at `http://www.deeplearningbook.org`.

Hand written digit recognizition using CNN

This is one of the classic "Hello World" type problem in the field of deep learning. We already covered one very simple case study of flower classification earlier and in this one we are going to classify hand written digits. For this case study we are using the `MNIST` dataset. The `MNIST` database of handwritten digits is available at http://yann.lecun.com/exdb/mnist/. It has a training set of 60,000 examples, and a test set of 10,000 examples. Some of the sample images in this dataset are as shown:

A typical hello world neural network that we are building is to train our network with the training set and to classify the images based on the test set. For this we will use a CNN or convolutional neural network.

A convolutional neural network is a special type of feed forward neural network and is especially suited for image classification. Explaining the entire concept of a convolution network is beyond scope of this chapter but we will explain it briefly.

A digital image on a computer comprises of many pixels . Using the intensity of these pixels in the various color formats for example RGB format we can represent an image in a mathematical matrix form. This matrix of nested arrays is called as a Tensor. Convoluntional networks consume these nested arrays of pixels or tensors and are able to extract features from it. These features are later fed to a multi neural network for further classification of these features. So, the ultimate aim of the convolution layers is automatic feature extracton from images based on weights and biases. For more information on convolutional neural networks refer to the book "Deep Learning" which we mentioned in the previous section.

We will be using DeepLearning4j for building this CNN and recognizing the handwritten digits. The next section will depict the code for this case study.

Diving into the code:

Before we look at the entire code we will get the setup ready in our IDE. For this example we are using `intellij` but feel free to use any IDE of your choice.

We will first create a maven project in intellij and we will use quickstart artefact for the project creation. After creating the project the project window would open inside intellij. Open the `pom.xml` file and add the dependencies, as shown:

```xml
<dependency>
   <groupId>org.deeplearning4j</groupId>
   <artifactId>deeplearning4j-core</artifactId>
   <version>0.8.0</version>
</dependency>

<dependency>
   <groupId>org.nd4j</groupId>
   <artifactId>nd4j-native-platform</artifactId>
   <version>0.8.0</version>
</dependency>

<dependency>
   <groupId>org.apache.spark</groupId>
   <artifactId>spark-core_2.11</artifactId>
   <version>2.1.0</version>
</dependency>

<dependency>
   <groupId>com.beust</groupId>
   <artifactId>jcommander</artifactId>
   <version>1.72</version>
</dependency>

<dependency>
   <groupId>org.deeplearning4j</groupId>
   <artifactId>dl4j-spark_2.11</artifactId>
   <version>0.8.0_spark_2</version>
</dependency>

<dependency>
   <groupId>org.nd4j</groupId>
   <artifactId>nd4j-kryo_2.11</artifactId>
   <version>0.8.0</version>
</dependency>
```

It is important to use the version of DeepLearning4j we have mentioned above, older DeepLearning4j libraries are not compatible with latest apache spark and can give an error in runtime.

Deep Learning Using Big Data

After your maven `pom.xml` file is ready, now create a java class, we will call it `LenetMnistSparkSample`.

Next, we create the two variables that we will be needed while running our code on spark

```
@Parameter(names = "-batchSizePerWorker", description = "Worker size")
private int batchSizePerWorker = 16;

@Parameter(names = "-numEpochs", description = "Number of epochs for training")
private int numEpochs = 15;
```

These parameters are explained in the table as shown:

Parameter Name	Description
batchSizePerWorker	This refers to the number of training examples per worker in Spark.
numEpochs	An epoch refers to one forward pass and one backward pass of all the training examples

We will next load the `SparkConf` object and pass the master as `local[*]`. We also set the application name. Finally we build the spark context using this configuration.

```
SparkConf sparkConf = new SparkConf();
        sparkConf.setMaster("local[*]");
            sparkConf.setAppName("DL4J Spark MLP Example");
JavaSparkContext sc = new JavaSparkContext(sparkConf)
```

DeepLearning4j out of the box provides a set of classes using which we can directly load these datasets from their web locations. We will use one of these classes called `MnistDataSetIterator` and pull the `mnist` dataset from its web location. As seen, we create two instances in one instance we build the training set and in the other instance we load the test set. The parameters passed to the constructor are the size of the batch on each worker of Spark, whether training is needed or not (we pass it as true) in this case the entire dataset of 60000 images is downloaded and used to train our network and finally we pass the seed value.

```
DataSetIterator iterTrain = new MnistDataSetIterator(batchSizePerWorker,
true, 12345);
DataSetIterator iterTest = new MnistDataSetIterator(batchSizePerWorker, true,
12345);
```

Next we convert the obtained dataset into a Java RDD of `dataset` objects. For this we first create an ArrayList and populate it with dataset objects from the `iterTrain` and `iterTest` instances. Next we use the Spark context parallelize method to load this data in memory. This is not the best way to do this as it would not work in case of huge datasets as it would load everthing in memory but for our small application it is ok to use here. This would create the Java RDDs of training and testing data.

```java
List<DataSet> trainDataList = new ArrayList<DataSet>();
List<DataSet> testDataList = new ArrayList<DataSet>();
while (iterTrain.hasNext()) {
    trainDataList.add(iterTrain.next());
}
while (iterTest.hasNext()) {
    testDataList.add(iterTest.next());
}

JavaRDD<DataSet> trainData = sc.parallelize(trainDataList);
JavaRDD<DataSet> testData = sc.parallelize(testDataList);
```

As mentioned earlier, for this case study we are using a convolutional neural network. One of the first convolutional networks that was made was "Lenet Network" and it was used to classify digits from the MNIST dataset only. We are going to use this simple Cnn network. As shown we create a simple `conf` instance and invoke the method `getLenetCnnConfig`. This method returns the `MultiLayerConfiguration` object that wraps the whole Cnn network within it.

```java
MultiLayerConfiguration conf = getLenetCnnConfig();
```

Let's see the code for this `getLenetCnnConfig`. At the start of the method we specify the local variables that are needed in this method.

```java
public static MultiLayerConfiguration getLenetCnnConfig() {
    int nChannels = 1;
    int outputNum = 10;
    int batchSize = 64;
    int nEpochs = 1;
    int iterations = 1;
    int seed = 123;
```

The attributes shown are

Variable Name	Description
nChannels	The number of input channels
outputNum	The number of possible outcomes
batchSize	The test batch size
nEpochs	The number of training epochs
iterations	The number of training iterations
seed	The seed used in shuffling the data

For each neural network, we have to specify the learning rate which the size of the variation made to the weights in each iteration. The smaller the value, the faster the algorithm trains, while with a larger values the algorithm takes longer to learn but performs better. We need to strike the right balance here. We specify the learning rate in a `HashMap` object with the number of iteration and the learning rate for that.

```
// learning rate schedule in the form of <Iteration #, Learning
Rate>
Map<Integer, Double> lrSchedule = new HashMap<Integer, Double>();
lrSchedule.put(0, 0.01);
lrSchedule.put(1000, 0.005);
lrSchedule.put(3000, 0.001);
```

After creating the variables and specifying the learning rate now is the time to build our convolutional neural network configuration. We start with the `NeuralNetConfiguration` class and invoke the builder method on it to start building our configuration. Next, we provide the parameters like seed and iterations as we depicted previously and we need to build the network as shown:

```
MultiLayerConfiguration conf = new NeuralNetConfiguration.Builder()
        .seed(seed)
        .iterations(iterations)
        .regularization(true).l2(0.0005)
        .learningRate(.01)
        .learningRateDecayPolicy(LearningRatePolicy.Schedule)
        .learningRateSchedule(lrSchedule)
        .weightInit(WeightInit.XAVIER)
        .optimizationAlgo(OptimizationAlgorithm.STOCHASTIC_
GRADIENT_DESCENT)
        .updater(Updater.NESTEROVS).momentum(0.9)
        .list()
        .layer(0, new ConvolutionLayer.Builder(5, 5)
        .nIn(nChannels)
        .stride(1, 1)
        .nOut(20)
```

```
                    .activation(Activation.IDENTITY)
                    .build())
                        .layer(1, new
    SubsamplingLayer.Builder(SubsamplingLayer.PoolingType.MAX)
                        .kernelSize(2,2)
                        .stride(2,2)
                        .build())
    .layer(2, new ConvolutionLayer.Builder(5, 5)
                        .stride(1, 1)
                        .nOut(50)
                        .activation(Activation.IDENTITY)
                        .build())
    .layer(3, new
        SubsamplingLayer.Builder(SubsamplingLayer.PoolingType.MAX)
                        .kernelSize(2,2)
                        .stride(2,2)
                        .build())
    .layer(4, new DenseLayer.Builder().activation(Activation.RELU)
                        .nOut(500).build())
    .layer(5, new
    OutputLayer.Builder(LossFunctions.LossFunction.NEGATIVEL
    OGLIKELIHOOD)
                        .nOut(outputNum)
                        .activation(Activation.SOFTMAX)
                        .build())            .setInputType(InputType.
    convolutionalFlat(28,28,1))    .backprop(true).
    pretrain(false).build();

    return conf;

        }
```

Let's go over some of the important parameters we depicted previously:

Parameter	Description
Optimization algorithm	This is the optimization algorithm that is used to optimize our weights and biases for our neural network and in our case we are using "Stochaistic Gradient Descent"
layer 0	A convolution layer
layer 1	A sub sampling layer
layer 2	Another convolution layer
layer 3	Another sub sampling layer

Parameter	Description
layer 4	A dense layer where a function like Relu is applied. Relu or Rectified linear unit is an activation function like Sigmoid. More information on this function can be found on this link: https://en.wikipedia.org/wiki/Rectifier_(neural_networks).
layer 5	This is the final layer where the output is received.
Back propagation	This specifies whether back propagation is used or not. It is used to find the minimum value of the error function that is used to calculate the error in prediction, this way we will get the best values for the weights and the biases. More information can be obtained on Wikipedia:https://en.wikipedia.org/wiki/Backpropagation
nOut	This specifies the output type. In our case it is a vector holding 10 values since we are predicting digits from 0 to 9.
Input Type	This is the type of input which is a nested array representing the pixels of the image in RGB format. Since the image is 28 by 28 in width and height and it is black and white only thus the 3rd dimension value is "1" and hence the array is of type [28,28,1].

For more detailed information on this configuration refer to the Deeplearning4j documentation at https://deeplearning4j.org.

Next we create a training master class instance and this is used to specify how the model would be trained across a cluster of machine (in our case it is multiple threads on a single machine) on spark. Here we specify batchSizePerWorker which is nothing but the number of samples used in training per worker or executor in Apache Spark.

```
TrainingMaster tm = new
        ParameterAveragingTrainingMaster.Builder(batchSizePerWorker)
            .averagingFrequency(5)
            .workerPrefetchNumBatches(2)
            .batchSizePerWorker(batchSizePerWorker)
            .build();
```

Next create the Spark network instance using the SparkContext, configuration and the training master instance.

```
SparkDl4jMultiLayer sparkNet = new SparkDl4jMultiLayer(sc, conf, tm);
```

Now create iteration over the number of epochs you had configured. For each iteration invoke the fit function on the training data this will make the neural network learn the training features from the training data. After completing we also print out that the epoch is completed.

```
for (int i = 0; i < numEpochs; i++) {
    sparkNet.fit(trainData);
}
```

Now test how well you have trained your neural network in recognizing the digits form the test dataset by running the evaluation on the test dataset. Deeplearning4j provides an option to store the results of this evaluation in an `Evaluation` object:

```
Evaluation evaluation = sparkNet.evaluate(testData);
```

Finally, we will print the result from the `evaluation` instance:

```
System.out.println("***** Evaluation *****");
System.out.println(evaluation.stats());
```

This would print the evaluation as:

```
***** Evaluation *****

Examples labeled as 0 classified by model as 0: 5922 times
Examples labeled as 0 classified by model as 8: 1 times
Examples labeled as 1 classified by model as 1: 6715 times
Examples labeled as 1 classified by model as 2: 3 times
Examples labeled as 1 classified by model as 4: 1 times
Examples labeled as 1 classified by model as 7: 19 times
Examples labeled as 1 classified by model as 8: 4 times
Examples labeled as 2 classified by model as 1: 1 times
Examples labeled as 2 classified by model as 2: 5945 times
Examples labeled as 2 classified by model as 4: 1 times
Examples labeled as 2 classified by model as 6: 1 times
Examples labeled as 2 classified by model as 7: 6 times
Examples labeled as 2 classified by model as 8: 4 times
Examples labeled as 3 classified by model as 2: 2 times
```

After printing how many times the digit from the test dataset was recognized correctly or not the evaluation instance would also print the accuracy of the model as shown:

```
========================Scores========================
Accuracy:        0.9973
Precision:       0.9974
Recall:          0.9974
F1 Score:        0.9974
======================================================
```

As seen the accuracy of the model is around 99% which is very good for a machine learning algorithm. This is also the key reason as to why deep learning has become so popular. For other values in the result above refer to the documentation on `https://deeplearning4j.org`.

With this, we come to an end of this sample case study using Deeplearning4j. Apart from this library, readers who are interested in researching deep learning further can check out the following books and resources as mentioned in the next section.

More information on deep learning

Deep learning is a vast topic and fast emerging as a new and upcoming research field. There are lots of free resources available on the internet that users can reach out to in order to increase their knowledge on this subject. Some of these resources are as follows:

- Geoffrey Hinton is a famous researcher and professor at the University of Toronto. He has made tremendous contributions to the field of artificial intelligence. He has given away one free course on AI on Coursera. It's a course we highly recommend.
- There are many other websites giving video courses on deep learning, such as *udemy* and *udacity*.
- Two professors from MIT have provided an excellent free deep learning book that can be read here: `http://www.deeplearningbook.org/`.
- There is also a free and easy-to-read book on neural networks available online. The book is available here: `http://neuralnetworksanddeeplearning.com/`.

These are just a handful of resources, but as this is a fast-growing field, it won't be difficult to find more resources online.

With this, we come to an end of our introductory chapter on deep learning.

This also brings us to an end of this book. Hopefully it should have been a fun journey for the readers as it was for me as a writer of this book. The book covered a lot of real life case studies and sample code and we wish and believe that it will serve as a good learning experience for our readers who are looking to make a career change or are already working in the field of big data and analytics. We also believe that this book would give an opportunity for the java developers who are new to the concept of big data and machine learning to get a good grasp of these technologies and take this knowledge and apply it in many real world projects.

Summary

This chapter gave a brief introduction to the field of deep learning for developers. We started with how an artificial neural network mimics the working of our own nervous system. We showed the basic unit of this artificial neural network, the perceptron. We showed how perceptrons can be used to depict logical functions and we later moved on to show their pitfalls. Later, we learnt how the perceptron's usage can be enhanced by making modifications to it, leading us to the artificial neuron, the sigmoid neuron. Further we also covered a sample case study for the classification of Iris flower species based on the features that were used to train our neural network. We also mentioned how the Java library Deeplearning4j includes many deep learning algorithms that can be integrated with Apache Spark on the Java big data stack. Finally, we provided readers with information on where they can find free resources to learn more.

Index

A

Activation Function 357
advanced visualization technique
  about 95
  IVTK Graph toolkit 96
  prefuse 95
Alternating Least Square (ALS) 263
Apache Kafka
  about 331, 332
  healthcare analytics 332
  IoT sensors, integration 332
  log analytics 333
  risk aggregation, in finance 333
  social media real-time analytics 332
Apache Spark
  about 12
  actions 14, 15
  actions, on RDDs 19, 20
  Apache Mahout 26
  Apriori algorithm,
      implementation 51-54
  common transformations,
      on Spark RDDs 18
  concepts 12, 13
  data, analyzing 17
  data, loading 16, 17
  data operations 17
  data, saving 23
  Deeplearning4j 26
  FP-Growth algorithm, executing 66-68
  machine learning modules 25
  paired RDDs 20, 21
  programs, executing on Hadoop 23, 24
  results, collecting 23
  results, printing 23
  samples, Java 8 used 16
  Spark Java API 15
  subprojects 24
  transformations 13
Apache Spark machine learning API
  about 125
  features handling tools 126
  machine learning algorithms 125
  model selection 126
  tuning tools 126
  utility methods 126
Apache Spark, machine learning modules
  machine learning libraries 25
  MLlib Java API 25
Apriori algorithm
  disadvantages 54
  implementation, in Apache Spark 51-54
  using 54
artificial neural network 355

B

bagging 213-215
bag of words 168
bar chart
  about 77-79
  dataset, creating 78, 79
base project setup 336
  default Kafka configurations, used 336, 337
  Maven Java project, for Spark
      Streaming 337-341
bayes theorem 157-159
bidirected graph 291
big data
  Analytical products 6
  Batch products 6

data analytics on 3
for data analytics 3, 4
Hadoop, basics 4-7
Machine learning libraries 6
NoSQL 7
Search 7
Streamlining 6
to bigger pay package, for Java
  developers 4
big data stack
Flume 7
HDFS 7
Impala 7
Kafka 7
MapReduce 7
Oozie 7
Spark 7
Sqoop 7
Yarn 7
binary classification dataset 112
boosting 215, 216
bootstrapping 213
box plots 88-95

C

charts
for data visualization and reporting 71
for initial data exploration 70
used, in big data analytics 70
clustering
about 268-270
biology 269
customer segmentation 269
data exploration 269
epidemic breakout zones, finding 269
for customer segmentation 280-287
hierarchical clustering 270, 271
K-means clustering 272, 273
k-means clustering, bisecting 273-275
news categorization 269
news, summarization 270
search engines 269
types 270
clustering algorithm
changing 287, 288

code
diving 374-382
cold start problem 248
collaborative recommendation systems
about 256, 257
advantages 257
collaborative filtering 258-266
disadvantages 258
common transformations, on Spark RDDs
Filter 18
FlatMap 18
Map 18
other transformations 19
Conditional FP Tree 64
Conditional Pattern 64
conditional probability 156, 157
content-based recommendation systems
about 242-244
collaborative recommendation systems 256
content-based recommender,
  on MovieLens dataset 249-256
dataset 248, 249
Euclidean Distance 244, 245
Pearson Correlation 246, 247
content-based recommender
on MovieLens dataset 249-256
context
building 34
customer segmentation
about 275
clustering 280-287

D

data
Apriori algorithm 43-47
average value, populating 32
basic analysis, with Spark SQL 33
cleaning 31, 201, 202
constant value, filling 32
converting, to proper format 33
discarding 31
formatting 122
Full Apriori algorithm 48
incomplete data, handling 31
loading 35

missing data, handling 31
munging 31, 201, 202
nearest neighbor approach 32
parsing 35
preparing 122
Spark SQL, for data exploration
   and analytics 43
Spark-SQL way 35-43
storing 122, 123
unwanted data, filtering 31
data analytics
 Apache Spark 12
 distributed computing, on Hadoop 8
 HDFS concepts 8
 on big data 3
data exploration
 about 197-200, 276-279
 of text data 169-174
dataframe 34
DataNode 10
dataset
 about 29, 30, 34, 168, 276
 airlines dataset 306
 airports dataset 305
 data 72
 data, munging 137
 fields 72
 full batch approach 365
 partial batch approach 365
 reference link 136
 routes dataset 305
 URL, for downloading 71
dataset, linear regression
 average price per zipcode, sorting by
    highest on top 138
 data, cleaning 137
 exploring 137
 linear regression model, executing 139-143
 linear regression model, testing 139-143
 number of rows 138
dataset, logistic regression
 categorical data 148
 data, cleaning 148
 data exploration 148-150
 data, missing 148
 data, munging 148

executing 150-153
testing 150-153
dataset object 205
datasets splitting
 features selected 191
 Gini Impurity 191-195
data transfer techniques
 Flume 120
 FTP 120
 HBase 121
 Hive 121
 Impala 121
 Kafka 121
data visualization
 charts, used in big data analytics 70
 with Java JFreeChart 69, 70
decision tree
 about 185-188
 advantages 195
 building 188-190
 data, cleaning 201, 202
 data exploration 197-200
 data, munging 201, 202
 dataset 196
 datasets splitting, features selected 191
 disadvantages 195
 for classification 186
 for regression 186
 model, testing 202-209
 model, training 202-208
deep learning
 about 353, 366
 accuracy 366
 advantages 366, 367
 information 382
 no feature engineering required 366
 use cases 366, 367
Deeplearning4j
 about 26, 373
 Avro 27
 data, compressing 26
 Parquet 27
 references 373
distributed computing
 on Hadoop 8

[387]

E

edges 290
efficient market basket analysis
    FP-Growth algorithm, used 54-60
ensembling
    about 212
    advantages 216
    averaging 213
    bagging 213-215
    boosting 215, 216
    disadvantages 216, 217
    Gradient boosted trees (GBTs) 219-221
    machine learning algorithm, used 213
    random forest 218
    types 213
    voting 212

F

feature selection
    backward elimination 118
    chi-square 116
    embedded method 118
    filter methods 115
    forward selection 118
    pearson correlation 115
    wrapper method 117
FP-Growth algorithm
    array items, by priority 57
    conditional patterns, from leaf node
        Diapers 62-65
    conditional patterns, mining 61, 62
    executing, on Apache Spark 66-68
    FP-Tree, building 57
    frequency of items, calculating 56
    frequent patterns, identifying from
        FP-Tree 61
    priority, assigning to items 56
    transaction dataset 56
    used, for efficient market basket
        analysis 54-60
Frequent Item sets 64
Frequent Pattern Mining
    reference link 66

Full Apriori algorithm
    about 48
    apriori implementation 49-51
    dataset 49

G

Gradient boosted trees (GBTs)
    about 217-221
    data exploration 222-230
    dataset, used 221, 222
    gradient boosted tree model, testing 236
    gradient boosted tree model, training 236
    issues, classifying 221, 222
    random forest model, testing 230-235
    random forest model, training 230-235
graph analytics
    about 298
    centrality analytics 299
    community analytics 299
    connectivity analytics 299
    datasets 305
    GraphFrames 300
    GraphFrames, used for building
        a graph 300-304
    on airports 304
    on flights 304
    on flights data 306-319
    path analytics 299
graphs
    adjacency list 292
    adjacency matrix 292
    common algorithms 294, 295
    common terminology 293
    plotting 295, 296
    refresher 290, 291
    representing 292, 293
graphs, common algorithms
    breadth first search 294
    depth first search 295
    dijkstra shortest path 295
    PageRank algorithm 295
graphs, common terminology
    degrees 294
    edges 293

indegrees 294
outdegrees 294
vertices 293
GraphStream library
 reference link 296

H

Hadoop
 basics 4-7
 distributed computing on 8
 features 5
 Hadoop core 8
 HDFS 8
Hadoop Distributed File System (HDFS)
 about 8, 326, 327
 data locality 9
 DataNode 10
 failover support 9
 fault tolerance 9
 Immense scalability, for amount of data 9
 NameNode 10
 Open Source 9
hand written digit recognizition
 using CNN 374
HBase 326
HDFS concepts
 about 8
 architecture 9
 components 10
 design 9
 simple commands 11
hierarchical clustering 270, 271
histogram
 about 80
 creating, JFreeChart used 81
 using 81
human neuron
 axom terminal 354
 cell body 354
 dendrite 354
hyperplane 88, 134

I

Impala
 advantages 326
 Apache Kafka 331, 332
 flight delay analysis 327-331
 Spark Streaming 333-335
 trending videos 341-346
 used, for real-time SQL queries 326
Iris dataset
 reference link 367
IVTK Graph toolkit
 about 96
 other libraries 96

J

JFreeChart API
 chart component, creating 79
 chart object, creating 74
 dataset loading, Apache Spark used 73
 dataset object, filling 79

K

K-means clustering
 about 272, 273
 bisecting 273-275

L

linear regression
 about 130, 131
 dataset 137
 used, for predicting house prices 136
 using 135, 136
line charts 82-84
logistic regression
 about 143
 dataset 147
 Gradient ascent or descent 145
 heart disease, predicting 146
 mathematical functions, used 144, 145
 Stochastic gradient descent 146
 used for 146

M

machine learning
 about 100
 analytics, executing on big data 119
 Apache Spark machine learning API 125
 at Netflix 100

categorical features 111
cross validation 111
data, obtaining in Hadoop 120
data, preparing in Hadoop 120
example 100, 102
features extracted from datasets 111-113
features, selecting to train models 114
Hand writing detection, on cheque
    submitted via ATMs 101
issues 107, 108, 109
model, selecting 110, 111
models, storing on big data 123, 124
models, training on big data 123, 124
numerical features 113
semi supervised learning 106
spam filter 101
supervised learning 102
supervised learning, case study 106, 107
text features 113
training/test set 110
type 102
un-supervised learning 104, 105
unsupervised learning, case study 106, 107
massive graphs
graph analytics 298, 299
graph analytics, on airports 304
on big data 297
maths stats
lower quartile 89
max 89
mean 89
median 89
min 89
outliers 90
upper quartile 90
mean squared error (MSE) 274
median value 91
MNIST database
reference link 374
model
selecting 123
storing 124
testing 202-209
training 124, 203-207
multi-layer perceptron
about 362, 363
accuracy 364-366

used, for flower species
    classification 367-372
multiple linear regression 134, 135

N

Naive Bayes algorithm
about 159, 160
advantages 160, 161
disadvantages 161
NameNode 10
Natural Language Processing (NLP) 113, 162
neural networks 354, 355, 356
N-grams
about 165
examples 165

O

OpenFlights airports database
reference link 305

P

paired RDDs
about 20, 21
transformations 21, 22
perceptron
about 356-359
issues 359
Logical AND 359
Logical OR 360
multi-layer perceptron 362, 363
sigmoid neuron 361, 362
PFP 66
prefuse
about 95
reference link 95

R

random forest 218
real-time analytics
about 322, 323
ad-processing 323
big data stack 324
fraud analytics 323
in healthcare 323

[390]

recommendations, giving to users 323
sensor data analysis (Internet
    of Things) 323
real-time data ingestion
    about 325
    Apache Flume 325
    Apache Kafka 325
    Cassandra 325
    HBase 325
real-time data processing
    about 325, 326
    Spark Streaming 325
    Storm 325
real-time SQL queries
    Apache Drill 324
    impala 324
    Impala, used 326, 327
    on big data 324
real-time storage 325
**Recency, Frequency,
    and Monetary (RFM) 275**
recommendation system
    about 240-242
    content-based recommendation
        systems 242, 243
    types 240-242
Resilient Distributed Dataset (RDD) 12, 34

S

scatter plots 84-88
sentimental analysis
    about 162
    bag of words 168
    concepts 162
    dataset 168
    N-grams 165
    on dataset 174-180
    stemming 164
    term frequency 165, 166
    Term Frequency and Inverse Document
        Frequency (TF-IDF) 166, 167
    term presence 165, 166
    text data, data exploration 169-174
    tokenization 163
sigmoid neuron 361, 362
simple linear regression 131-134

smoothing factor 161
SOLR 326
SPAM Detector Model 102
SparkConf
    building 34
Spark ML 125
Spark SQL
    context, building 34
    dataframe 34
    data, loading 35
    data, parsing 35
    datasets 34
    SparkConf, building 34
    used, for basic analysis on data 33
Spark Streaming
    about 333-335
    base project setup 336
    cumulative calculations 335
    data collection, in real time 335
    predictive analytics, in real time 335
    storage, in real time 335
    use cases 335
    windowed calculations 335
stemming 164
stop words removal 163
Storm 334
sum of mean squared errors (SMEs) 273
supervised learning
    about 102
    classification 104
    regression 104
Support Vector Machine (SVM) 181-183

T

tendency 246
term frequency
    about 166
    example 166
**Term Frequency and Inverse Document
    Frequency (TF-IDF) 166**
    about 167
    inverse document frequency 167
    term frequency 167
TimeSeries chart
    about 71
    all india seasonal 71, 72

annual average temperature series
  dataset 71, 72
multiple TimeSeries, on single chart
  window 75, 76
simple single TimeSeries chart 72-74
tokenization
  about 163
  pre-trained model, used 163
  regular expression, used 163
  stop words removal 163
trending videos
  about 341-346
  sentiment analysis, at real time 346-351

V

vertexes 290
Visualization ToolKit (VTK)
  about 96
  URL 96

W

windowed calculations 342